Selfie

James Sherry

Selfie

Poetry, Social Change & Ecological Connection

palgrave
macmillan

James Sherry
New York, NY, USA

ISBN 978-981-19-4869-5 ISBN 978-981-19-4870-1 (eBook)
https://doi.org/10.1007/978-981-19-4870-1

Cover illustration: Maram_shutterstock.com

This Palgrave Macmillan imprint is published by the registered company Springer Nature Singapore Pte Ltd.
The registered company address is: 152 Beach Road, #21-01/04 Gateway East, Singapore 189721, Singapore

Acknowledgments

Thanks to Deborah Thomas for her support, Jonathan Skinner for his patience, and Joshua Schuster for his ongoing encouragement.

"Borderlands: Am I I & They? Knowing Without Or" (2020). Was published in *On Air*. An online project by Carico Massimo (IT). Italian translation by Viola Cateni. Available at: https://on-air.caricomassimo.org/en/airquakes/contents/borderlands-am-i-i-they-knowing-without-or-from-selfie-poetry-ecology#content (Posted: 11/20/2020).

A summary of "How Can Culture Change Habitat?" appeared in Jacket2 with the title "Metaphor and social media in the 2020 election" in a series curated by Michael Ruby and Sam Truitt. https://jacket2.org/commentary/metaphor-and-social-media-2020-election, 10/29/2020.

Two poetry segments from this book were published in *The Millbrook Independent* edited by Kevin T McEneaney under the titles "B&E" https://themillbrookindependent.com/?p=1698 on 11/30/2021 and "Migrant" https://themillbrookindependent.com/?p=1656 on 12/07/2021.

A summary of the different discussions in Selfie about scale titled "Scalar Properties" is forthcoming in Paideuma, "Poems & Places" in fall 2023.

ACKNOWLEDGMENTS

Contents

List of Figures

Identity's Constructive & Connective Ecosystem

Introduction

Paul Valéry said, poetry is a "language within a language." Poetry is not, however, a separate language any more than words are easily separated from each other.

> Surely you have noticed the curious fact that some *word* which is perfectly clear when you hear it or use it in current language, and causes no trouble when it is caught in the quick pace of an ordinary phrase, becomes magically embarrassing, introduces a strange resistance, and foils all efforts of definition as soon as you take it out of circulation to examine it apart, and seek a meaning in it after having re- moved it from its momentary function?[1]

Multiple definitions of the same word accept some responsibility for Valéry's difficulties. Meanings intersect and different words have similar meanings. The history of words, as distinct from current usage, adds complexity to understanding individual words. And through their relationships to other words like position in a sequence, part of speech, and logic in a phrase, words function more effectively in a variety of uses from poetry to public relations to recipes.

Words together with spaces and punctuation form poems. The choice of words and their order, informed by authorial intention, past events, immediate sensations, ideas, and a host of prior constructions of language, emerge from sources like other poems, daily speech, and the process of writing. An immense diversity of poetic components and possibilities

J. Sherry, *Selfie*, https://doi.org/10.1007/978-981-19-4870-1_1

encourages writers to limit the mechanisms they use for comprehensibility, to assert their style, and to seek common ground with a group of peers. Ezra Pound's famous equation, "DICHTEN = CONDENSARE,"[2] may have been an aesthetic decree, but there are other environmental, biological forces and explanations propelling individual and group tendencies in the same direction.

At the same time, life expands and so does poetry. Assemblages of poetry, composed of assorted poems, poets, methods, mechanisms, and tendencies of writing, operate together in concert and in conflict. Their processes connect through an ecological network, a phrase that is not only a figure of speech but also demonstrates how other scales of life in the biosphere operate linguistically as language functions ecologically.

Poetry is derived from, nourishes, inflects, and infects other language structures in one's self, society, and surroundings. Poetry is part of, confused with, overlapping, and yet distinguishable from other language acts. Models of poetry inhabit larger models to shape how people use language for meaning and joy, to inform thought, and to spur action. That ecology scales down to poetry, and poetic models scale up somewhat.

By reading poetry through these lists of the multiple identities of poets and poems, I intend to focus less on poetry as an autonomous art and more on how it relates to, reacts with, and intersects other processes in the biosphere. For now, enough has been written on uniqueness. Although poetry can be fruitfully considered independently, in *Selfie* I approach it more as an ecological process connected to other writing, other uses of language, other disciplines of thought, and social and environmental activism. Through these channels, poetry can contribute to an apparatus of change for rethinking one's self, social relations, and living interactions with one's surroundings.

Not all writers consider poetry as active, connective, and influential. Instead, many poets treat poetry as an isolated, genuine, and self-governing practice, a kind of independent organism. The poet as outcast appears famously in Samuel Taylor Coleridge's "Rime of the Ancient Mariner." Edgar Allan Poe appeared for Charles Baudelaire as the image of isolation: "the United States was for Poe only a vast prison through which he ran, hither and thither, with the feverish agitation of a being created to breathe in a purer world."[3] Then, consider the solitary image of the Fascist Pound locked in a cage in Pisa.

Yet despite their temperamental differences, John Keats and Percy Shelley spent time together in 1817. Pound acted as secretary to William Butler Yeats. John Ashbery and James Schuyler wrote *A Nest of Ninnies* together. The Beats traveled together in *On the Road* by Jack Kerouac. To this day, Saint Mark's Poetry Project acts as a gathering place for poets of the eponymous New York School. Over centuries, Asian poets developed widely used collaborative forms. In the late 1970s, five language writers—Bruce Andrews, Charles Bernstein, Ray DiPalma, Steve McCaffery, and Ron Silliman—collaborated in their signature poem cycle, *Legend*. In line with other multiples, poets often write in series: Dante Alighieri's *La Vita Nuova*, Rainier Maria Rilke's *Sonnets to Orpheus*, Zhai Yongming's "Woman," and Bob Perelman's *Jack and Jill in Troy* make it clear that poems are often written together.

At the level of understanding, poetry famously inscribes multiple meanings. Even W. H. Auden's famous phrase, "Poetry makes nothing happen," has at least two separate references: nothing as ineffectual, that is, remaining within poetry, and nothing as invoking the void, a larger frame than daily traffic. In the same stanza of Auden's poem "In Memory of W.B. Yeats," poetry also "flows," "it survives, / a way of happening."[4] These are not nothings, they are actions, what poetry does, one initialed poet to another.

Adding to Auden's poetic activities, William Wordsworth's poet recollects after the event.[5] Wallace Stevens' poetry "must resist the intelligence / Almost successfully," and resist identity, "A brune figure in winter evening resists / Identity."[6] For Gwendolyn Brooks poetry "Is Life Distilled."[7] perhaps from Emily Dickinson's "Distills amazing sense."[8] For Lyn Hejinian poetry rejects closure.[9] For Will Alexander the poet scatters criteria and engenders gnostic recognition.[10] These diverse identities and actions of the poet and poem connect in an ecology of language that each poet creates, assumes, and shares with their group of readers and other writers. Some poems flicker and some are concrete. Readers might like some poems and styles better. Readers may think that certain times call for a particular poetics; some may think poetry is eternal. As bell hooks says: "Poetry sustains life. Of this I am certain."[11]

In *Selfie,* I will suggest that through these many linked actions and conditions of poets and poems, poetry changes how we think about language, ourselves, our societies, and our surroundings. Sometimes poets understand their own poetry's scope, context, and impact while other poets focus closely on the poem and ignore or even avoid peripheral concerns. In *Selfie*, I shape an awareness of how poetry can contribute to environmental conditions, in addition to individual thought and social concerns.

But why and how would changes in language help slow global warming? How can poetry read by so few people affect the monstrous scale of global processes? I want to discuss that incongruity in this book, not to solve a problem, but to show the connections, scaling, and similarity between processes like poetry and climate, to build an extensible model.

Although people see and know the future danger of climate change, they are often compelled to act on more immediate concerns such as small pleasures, daily threats to their agency, social benefits to their groups, and methods to gain near-term objectives for family, government, profit, thought processes, and poetry. These intervening demands of lives, careers, and social groups tend to dominate the more distant and difficult-to-imagine changes from global warming. Examining the actual and immediate causes of climate change helps clarify this struggle between near-term challenges and distant threats.

Emissions of greenhouse gases from six industries—transportation, energy, manufacturing, construction, mining, and agriculture—cause global warming. Industrial processes are linked to systems in the biosphere. *The Climate Report*: "These interconnected systems are increasingly vulnerable to cascading impacts that are often difficult to predict…"[12] Getting ahead of the problem pays: "Every $1 spend on preparation saves society an average of $4" later on.[13] For decision makers in control of these processes, the incentive to reduce emissions will not easily emerge from the current torrent of global expansion and unquestioned humanist assumptions. Immediate self-interest prevents owners and influencers from actually changing their behavior even when they acknowledge the risks of global warming. Yet obscuring these connections make change more difficult and costly.

Changing these complex apparatuses that civilization built to protect the lives and limbs of citizens poses further difficulty for leadership. No one person or group controls the complex supply chains that drive emissions. Only concerted action can plan effective change. Residual biases push the politicians of many large countries toward passively accepting spewing pollutants into our air in order to retain jobs and security. Biases include consumer narcissism, private surplus, financialized production, competitive self-interest, models of nature as too large and perfect to be affected by mere mortals as the optimum principles of social interaction, and language constructions that separate and foment argument instead of understanding multiple points of view as components of an ecosystem of ideas. The items in this list of biases are usually considered separately to deflect individual leaders from making the sacrifices and combinations necessary to slow climate change.

Individuals, citizens, and consumers in both capitalist and socialist ecosystems have difficulty changing their habits around plastic and petroleum without help from language renewal, new narratives, and meaningful regulation. While developed nations will need to implement wrenching changes in how their people live day to day, digital technology and social media sensitize users to interpret even minor alterations in habits, facts, and attitudes as threats to their stability and agency. I don't propose people change their minds by reading a poem, but poetry warping language and creating new phrasing does participate and acts as a heuristic for climate change. As Henry David Thoreau opined:

O Nature! I do not aspire
To be the highest in thy quire,—
To be a meteor in the sky,
Or comet that may range on high.

But those aspiring to influence events cannot hide as he hoped in the next quatrain:

Only a zephyr that may blow
Among the reeds by the river low;
Give me thy most privy place
Where to run my airy race.[14]

Most wealthy people will not change how they think about and treat their surroundings without significant pressure being placed on their self-image, their group affiliations—family, community, ideas, and supply chains—and their assumptions about the availability of resources needed to secure their safety and freedom to pursue self-interest. Appeals by science to facts about climate change have only convinced some people and appear insufficient by themselves to force wholesale policy change in the time we have left to avoid the collapse of many societies.

Fortunately, poetry explores, in both incisive and fuzzy detail, the questions of scalability in the biosphere, how one thing stands for another, creating real connections between things as physical, mental, ethical, and linguistic similarities and equivalents. Metaphor, one of poetry's methods, dictates many of our assumptions, explored in the chapter "Networks of Metaphor." At another level, we know how to change these views of self, society, and surroundings; we've done it before by refocusing language and distributing messages, concepts, and aspirations through channels that influence opinion. Americans changed how we thought about ourselves after World War I and World War II and again after the Viet Nam War and again after 9/11, demonstrating that changes in group identity occur frequently, and public information and novel language support those changes. Europe, Africa, and Asia have gone through similar adjustments in identity. At both the distilled level of poetry and the practical level of public relations, minds can be prepared, encouraged, and propelled to change by altering linguistic ecosystems.

Russian intellectuals and poets have changed their views of self and state after the Revolution, again since the death of V.I. Lenin, and again since the collapse of the Soviet Union. Chinese women, especially the poets I know among Third Generation writers and Original Poets (who prefer to remain anonymous), tell me that they think differently about themselves and their agency today than their grandmothers did before the Revolution. The rising classes in India now look beyond its shores to an expanded set of values as they assume roles in the global technology culture. A person living *in* society and their self-image formed *by* society are very different from those of my youth just after World War II. We can change the assumptions of people and even of leaders by changing language and relationships through specific and documented connections.

Methods used to implement change have an eerie similarity. George Creel and Woodrow Wilson worked with the Committee on Public Information in World War I to change how Europeans viewed America. Edward Bernays convinced consumers to accept fluoride in drinking water and dozens of other products through *The Engineering of Consent*, the title of his 1955 book.[15] And who was not more careful with campfires after Smokey the Bear said, "Only *you*…"? A trusted general convinced America that Iraq was about to use weapons of mass destruction that did not exist. Some of these campaigns were true, some false, and some redirected citizens' attention toward where leadership wanted them to focus—all worked through "informing" (and conforming) the public view.

In a specific example, the term "climate change" was toned down from "climactic change" in its original use in 1956. Frank Luntz, the Republican pollster and strategist, worked to replace the inflammatory term "global warming" with the neutral term "climate change," saying "'climate change' suggests a more controllable and less emotional challenge," even as he and his cohorts sought to avoid scientific consensus.[16]

The public relations professionals who promote methods to slow climate change need impetus from informed citizens, financing from willing corporations, understanding of scalar complexities from science, scalable language innovations from poetry, and leadership to execute consistent programs of the size and extent required to adapt corporate mentality in the huge populations of nations such as the US, China, Russia, Japan, and India, the chief contributors to climate change. Underlying changes in societies and cultures will inevitably materialize as the environment degrades, but it would be useful, and, without understatement, one might say urgent, to begin alterations before they are enforced by social collapse.

To accomplish similar changes in support of green initiatives, underlying concepts of how individual agency operates, social structure evolves with habitat, and human/planetary hierarchies interact through language must be adapted for considerable and urgent consumption. *Selfie: Poetry, Social Change & Ecological Connection* demonstrates one reasonably complete model, how such an adaptation might work through rethinking poetry, what poetry does, and how poetry does it. Alternative grammars, reconceiving metaphor, and adapting form increase our awareness of a non-singular way of using language, scaling up to how people operate with the world and world operates with us.

Poetry already provides tools and methods to present the complexities of non-linear climate systems to readers in a way that has been difficult for biologists and climate scientists to communicate with facts. I want to focus on language writing[17] and the group of experimental and avant garde poetry rather than ecopoetry[18] because language writing tends, with notable exceptions like Jed Rasula's *This Compost*[19] and Caroline Bergvall's *Drift*,[20] Cecilia Vicuña's language performances in many parts of the world,[21] and Bob Grenier's "Northern New Hampshire"[22] to include both personal perspectives and the relations between individuals and society—a limit that most political leadership shares. Addressing the inherent and often unintended connections of language-centered writing to its ecosystems deals directly with the problem of expanding readers' outlook to include our surroundings. Ecopoetry and ecopoetics are more directly and transparently concerned with environment. For ecopoetics, the problem of changing perspectives seems less urgent and at the same time more difficult. Many poets focused on "nature" already use non-descriptive writing.

Language-centered writing and some other experimental strategies use alternative structures that might change individual and group biases. Beyond that poetry can be an effective agent of change through the memorable phrases in the verses of popular music, the descriptions of ecopoetry, activist verses from Thoreau through contemporary Native American ecopoets like Tommy Pico, and evocative verses by the millions. These musical verses and poems have largely asserted a love of nature and reiterated nature's separation from humanity.

The environmental model in *Selfie* proposes that poetries and poetics are built toward multiple perspectives. Just as diverse individual identities, group affiliations, and habitats support life, so varieties of style and method support poetry. Understanding that there are many poetries and that each poetry affects readers differently at different levels begins to change the ideological boundaries of poetry and even the very characteristics of those boundaries. Poetry can be applied to the ways people conceptualize and act on climate change through both specialized and daily use of language. Readers, writers, publishers, and distributors of poetry already expect multi-level perspectives from poetry in the way components of a metaphor (and figures of speech, like synecdoche) perform their roles by linking two or more domains, scales, and dimensions. Can we extend that expectation of metaphoric linkage to other domains, like climate, so readers can expect and accept multiple meanings, solutions, and ways of living in their surroundings?

I titled this book *Selfie* because I realize that I am building its multidimensional model from my own point view as well as making connective sense. Tellingly, the background of any selfie is often the reason for taking the photo then and there. In the background, you'll see where I am, who I'm with, how I adapt my perspective to different locations, and adopt other relevant writers. I want to show the background in different lights that reveal even a single habitat as multiple. I try to remain cautious about myself recognizing that many selfies obscure the background by making the self-image too large in the frame. I hope my discretion in these matters will not be construed as lack of commitment to my argument or the collective poetry.

I have already written on climate change and poetry, starting with *Our Nuclear Heritage*, a loose narrative, poetry, and essay text published in the early 1990s, then *Oops!*, mostly essays with poetry, followed by *Entangled Bank*, poetry built around the environmental model, and most recently *The Oligarch*, a rewriting of Niccolò Machiavelli's *The Prince*, to show the difference between form and operations of governance in political ecosystems.

I also support waiting until age begins to obscure memory before writing any autobiography. Performing this objective is another reason I focus on the language-centered writing that I have been immersed in during 40 years of writing, editing, and publishing, hence the plethora of citations from Roof Books where I have been the primary editor. It's what I know and where I've been, a hybrid of my own and academic discourses, a selfie in the context of poetry.

The specialized ways poetry acts on readers, as an assemblage of individual insights, public language, and non-fiction, appear to be direct, indirect, and at multiple levels rather than telling us what to do like a sermon. So, a more accurate picture of how poetry might affect the mentalities associated with climate change can be gained from looking at the complex sets of connections than from descriptions of self, society, and location isolated from each other. This multi-level model will tend to be more accurate than the easily inferred, charming stories about what happened to me. But to be thorough and because narrative operates through any sequence, there are several places in *Selfie*'s sequence of events where I recount my own tales and allow others a voice to tell as well.

I'm thinking at the moment about diversions in the *Decameron* since I'm writing this introduction in the middle of the Covid-19 plague. But *Selfie* is not intended as a distraction or even only documentation, a "photograph that one has taken of oneself, *esp.* one taken with a smartphone or webcam and shared via social media."[23] *Selfie* is intended to be a series of grammatic and rhetorical examples connected across multiple contexts, a model. I hope you'll read this book not so much as an argument for one way of looking at, interacting with, and being shaped by environment, but as a many-to-many set of methods demonstrating how multi-level perspectives build connections between and within selves, to others, and to shared surroundings differently than usually expected.

Although *Selfie* performs various versions of my own identity, I would not assume that I, as an individual, can have much immediate impact. Most environmental reformers discount the effect of the individual on the planetary environment, preferring to work on industrial, political, and conceptual processes. The popular press, however, being an industry itself, often ignores how industry affects climate change. Instead, corporate media makes lists of things that you, the all-important, consuming individual, can do like power your home with renewable energy, maintain your vehicles, recycle your trash, and not waste the food you buy.

Although there's nothing wrong with recycling, these lists steer you back to yourself instead of focusing outward. Focusing outward, not merely inward toward self-knowledge, is one of the key skills individuals can develop in support of environmentalism. To avoid naïve dualism, I point out that good understanding of oneself improves the accuracy of perceptions of externalities.

At another level, in the regions that cause the most pollution like the US, both mass cultures, with their focus on sentiment, and high cultures, with their focus on how society affects individuals and aesthetics, encourage writers, readers, and politically aware citizens to concentrate on themselves and their peer groups for answers to social problems. Mass and high culture tend to drive people's attention inward, which is also how people tend to respond to uncertainty. I am writing *Selfie* as a way to clarify the actions between selves and societies by adding ecosystems to propel a more thorough comparison than psychology and social critique as a liberal pair.

Meanwhile, almost imperceptibly, yet widely discussed in the news media, the pressures of external threats as diverse as climate change, autocracy, and illness overwhelm and victimize writers who continue to focus on themselves and their groups. Studies of the impact of stress on individuals produce some results I didn't expect. In their article "Acute social and physical stress interact to influence social behavior: The role of social anxiety," Bernadette von Dawans et al. remark:

> We found significant influences on various subjective increases in stress by physical and social stress, but no interaction effect. Cortisol was significantly increased by physical stress, and the heart rate was modulated by physical and social stress as well as their combination. Social anxiety modulated the subjective stress response but not the cortisol or heart rate

> response. With respect to behavior, our results show that social and physical stress interacted to modulate trust, trustworthiness, and sharing. While social stress and physical stress alone reduced prosocial behavior, a combination of the two stressor modalities could restore prosociality. Social stress alone reduced nonsocial risk behavior regardless of physical stress. Social anxiety was associated with higher subjective stress responses and higher levels of trust.[24]

Selfie concentrates on balancing individual, social, and ecological interactions and identifying connections, that is, the paths of interaction between them. In this way, language, bodies, societies, and ecosystems are framed to look like more like each other with ecosystems writ at multiple scales supporting social interactions and multiple individual identities in fluctuating surroundings.

From movies to poetry, contemporary art frequently promotes resistance as the best way to foment change. (Remember the Wallace Stevens quotation above?) The environmental model maintains that, although resistance is important to climate change, its power operates effectively only in concert with building alternative agendas. Otherwise, resistance can appear childish or narcissistic. Individual and social resistance to polluting power, implementing new methods to reduce pollution, and developing the skill to understand and absorb other points of view together build the complex entity sufficient to change production methods and reduce reliance on polluting and mono-logical systems.

The environmental model predicts that as climate adjusts individuals and groups will begin to feel differently about themselves, write and act differently, adapting to change like most organisms. *Selfie* assumes that individuals and social groups will adapt future selves increasingly around climate stresses. Inversely, variations in collective behavior will influence climate as they have already. Poetry moves individuals and poetry is changed by social and environmental forces. The types and quantities of food available to eat, the weather, what we see and hear from friends, the kinds of houses we live in, and what public and social media will say about us are all impacted by local climate conditions including the climate of public discourse in specialties like poetry and the climate of individual thought. We accept this generalization because as Raymond Holder Wheeler says climate in different parts of the world significantly affects how we interact and build our cultures.[25]

The Australian Academy of Sciences points out that humans have lived in a relatively stable climate for the past eight to 12,000 years.[26] Now, as industry rapidly changes climate, uncertainty and conflict among neighboring individuals and groups increase. Humans, according to Sabine Perch-Nielsen et. al., begin to move about more, seeking security.[27] Even as we need collective action, the stresses of climate change make our differences appear more threatening due to threats to our bodies and the availability of resources to sustain us. Compare global capitalists' well-funded fantasy of escape to Mars, families who cannot easily leave their small city with its contaminated water supply (Flint, Michigan is a well-publicized example), and worldwide reactions to desperate migrations from Central America, Africa, and the Middle East documented in a White House Report on migration.[28]

Capitalism drives human expansion and exploitation of environment without renewal, but at the same time environment shapes capitalism. As uncertainty increases, negative reactions to other people amplify while unscrupulous politicians exploit this distrust of difference to gain power for further extraction of wealth by the oligarchs that support their regimes. All this occurs at the same time that we need coordinated action and cultural change to manage industries and ways of working. Yet the most effective structure for initiating change, small group action, means oligarchy also promotes change even as an oligarch's primary concern remains to stay in power. Writing and activism around climate can adapt attitudes of leaders to our surroundings, but it's not an efficient process. Projecting and working with a model that includes habitat, social interaction, and individual behavior with their specific connections helps people adapt:

Adopt to Adapt
We can adopt
and then adapt.
We are adept at that.

Many components of our selves contribute to our ability to adapt: genetic, epigenetic, event, memory, posture, and feedback mechanisms within individuals as well as complex connections with society and habitat including observation, those second-order dynamics. The model characterizes these components' inter- and intra-action together as environmental, intending to reinscribe the ways humanism has negotiated

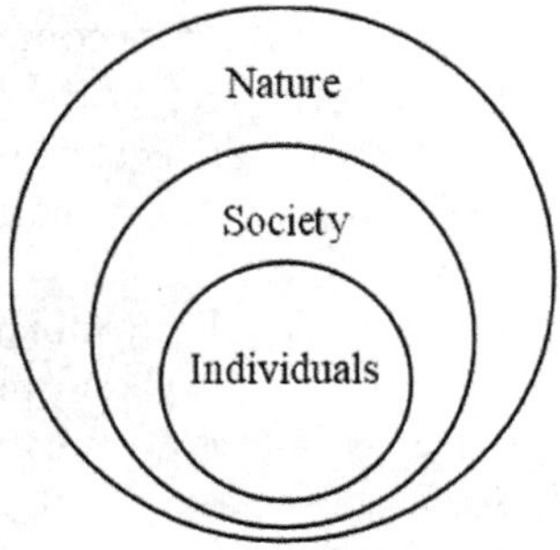

Fig. 1 Humanist model of individuals, society & nature

multiple individual points of view about what's me and not me, self and non-self. Humanism has represented selves, societies, and habitats as a series of concentric circles with the individual at the core, society in the middle, surrounded by nature as a location (Fig. 1).

Both Platonism (where humans see distorted cave shadows while the bright sun of nature reveals the truth that we "will not be able to see anything at all of what are now called realities"[29]) and some Buddhism (where critiques of the society's goals appeal to idealized meditation on self in nature to illuminate human illusions) separate humanity and nature in concentric regions.

In contrast to nature as ideal and separate from distorted human views, the environmental model proposes that natural processes are as inefficient and indeterminate as human minds and cultures. As Charles Darwin and contemporaries like paleontologist Stephen Jay Gould point out, organisms and processes in the biosphere need only to function in an ongoing manner, because organisms derive from a common ancestor.

> Evolution is the conviction that organisms developed their current forms by an extended history of continual transformation, and that ties of genealogy bind all living things into one nexus[30]

Organisms do not have to be perfect, they are not perfect, and they will not become perfect. As climate changes under our noses, appeals to the perfection of nature should be readily perceived as deceptive, even as we look to biomimicry for some solutions to global warming. Humans cannot become perfect on these paths and functioning in an ongoing manner appears desirable. Poems cannot become perfect and an effective critique of poetry aligns with the rest of the biosphere by accepting ranges of possibilities rather than the idée fixe of perfection. Each step toward the ideal must be real and therefore tentative, able to be changed as editing words in a poem.

Climate change is not only a capitalist problem, but a result of the human desire to make life easier, less risky, and more controllable. Both socialism and capitalism strive to improve human welfare by exploiting natural resources. Public social exploitation and private capital exploitation result in similar pollution patterns in China, India, and US. When socialists focus on self and society, they easily lose sight of habitats. When capitalists focus solely on personal well-being, they decrease their understanding of the needs of others and their impact on our surroundings. Optimistic action with others through careful review of our minds, our collective needs, and our habitats increases compassion.

To solve these problems caused by modern industry, some people support retreating to values of the past. Although traditional cultures can be useful as an example in addressing climate change, no single past model remembered as ideal can integrate all the contemporary complexities of diverse and dynamic individual psyches, defended group identities, huge populations, and multiple ways that people look at and are constituted by their surroundings.

As organisms and selves, we continue to attempt to reposition the world from a single perspective, yours in your case and mine in mine. I can feel myself wanting to simplify this text toward a singularity as I write these sentences. Linear reading of sentences pushes me to find a specific thought. Paragraph structure pushes me in the direction of finding myself in what I link together. As Gertrude Stein says in her essay "Poetry and Grammar" that shaped *Selfie*'s early chapters, "Sentences are not emotional but paragraphs are."[31] These categories push me further that direction of paragraphs as well and require rethinking how they interact.

Looking at the world from multiple perspectives means that practical and theoretical solutions can only be accurate within quite wide ranges as is clear when reviewing any historical system. Accepting this rangy reality remains difficult for even well-disciplined people. The knotty conflicts of self-interest do not disappear when expanding the purpose of and writing toward decentering humanity in our model of the biosphere.

As a group of civilizations sharing the same biosphere, we no longer have time to wait for everyone to think the same way, although increasing autocracy around the world pursues regional homogenization. We humans will benefit from increasing the speed of change toward sustainable cultures to match the speed of climate change. We will benefit from identifying with policies that include awareness of our surroundings. To do so, we can absorb the social value of diversity and ecological connectivity from our habitats. Put another way, I and we and even they must find ways to act not only through personal and social impulses but also to refract desire through and in response to place/location/situation. We are always in multiple, simultaneous states of being. Inversely, climate structures frequently scale even as scalar discontinuities persist. In fact, scale implies gaps in size and dimension as one scale is distinguished from another. At the same time, scales are linked by comparison.

We are connected differently
and multiply;
places, too, are changing.

Poetry has changed in similar ways for millennia with styles succeeding one another, thus poets and other people, too, must already be good, if noisy, at adapting. That noise might be politically or genetically dialectical. New poems are criticized by older poets and critics until they die or stop being read by a new generation and new ideas are accepted. But culture for the most part has not acknowledged multiplicity except through politics to gain control of marginal groups' votes.

Instead, many influential people promote the delusion of an inviolable unity of identities called self to mirror and reinforce how our organisms protect our bodies. Practically, both animal and plant bodies are generally denser than the air around them. Our vision of ourselves is distorted in some ways as a matter of survival. At the same time, we imagine that nonhuman activities operate with pristine perfection. Let us actively seek to comprehend and act on the fragility of our surroundings rather than have understanding thrust upon us.

The environmental person that the model proposes is constructed in Group 1 through language and parts of speech, since our definitions of ourselves appear in the words, concepts, and images that we identify with. To be honest as a writer, I acknowledge my starting point and my motivation to understand the people around me. Continuing from Group 1's pronouns and articles, nouns and verbs, the particles and building blocks of the social identities of class, ability, gender, and race, Group 2 moves to investigate syntax, metaphor, and the social values of flows associated with scale and probability. In Group 3, environmental social structures and inclusive hierarchy add potential practices to environmental models.

Throughout I have attempted to undermine any single reading of the text and intention. I try to avoid Aristotelian argument to confirm a single hypothesis, although I'm sure linearity creeps in as explanation. Some readers have complained that this technique makes the book difficult to read. But the difference in possibilities between the individual/society binary and the four-part structure of self, society, environment, and connection makes the problem both harder to describe in linear sentences and more likely to support climate change remediation.

In the chapters that follow, I have written in both poetry and prose and imported the language of others. I have used images and matrices to encourage multiple readings and reading methods. I keep telling myself to apologize for so many generalizations. In these ways, the page can be read in multiple, material dimensions—each word sinks and rises, held less tightly to its plane—so the effect of multiple perspectives is instantiated in the text.

Farmers know how to use the methods of plants and animals to grow food rather than implementing demonstrable falsehoods that support profits for landowners while sterilizing the soil. Poets know how to use the methods they find in the language as it structures the biosphere. Yet many revert to the partial representations of individual perception and social critique. The new generation of poets will compromise plain truth when the unvarnished facts will harm people, as long as we can attribute minimal damage done through the spin. But when the falsehood generates too much damage, we repair our strategy and take risks. In uncertain times, judgment becomes more problematic. To help address this conflict, the environmental model builds inconsistencies into actions and words. Only readoption of continually updated models and adaptation to our sites of action make life on a dynamic planet possible. There is no desirable end point.

Multidirectional Writing

"I have to be affected to say 'I' at all." *Judith Butler*

"I didn't mean to become an I." *Trish Salah*

The Particles

This chapter introduces how *Selfie* understands the relationships between individuals, social groups, and ecosystems by proposing physical and conceptual connectors both within and between each component of the biosphere. Particles of language such as pronouns and articles build multiple identities through examples from second wave feminism and twenty-first-century identity politics. Environmental identity willingly breaks existing grammatical agreement to increase social and environmental cohesion. Scale is a key relationship for climate change and invokes similarity as a third term in the binary of equality. Similarity works to bind ecosystems together. Finally, the chapter shows a diagram of the four-part environmental model of self and begins to derive how the model is built in accordance with a few historical and contemporary poetries.

* * *

> For a long time I have hesitated to write a book on women. The subject is irritating, especially to women; and it is not new. Enough ink has been spilled in the quarrelling over feminism, now practically over, and perhaps we should say no more about it.[1]

These sentences read oddly in retrospect since it appears Simone de Beauvoir's "subject" will not go away.

J. Sherry, *Selfie*, https://doi.org/10.1007/978-981-19-4870-1_2

Pronouns played a big role in the 1970s. I remember struggling to rewrite my speech during those years. I stopped using *his* and *him* as defaults, tried not to balk at being corrected when my habits of speech took over, and followed the discussion of alternative constructions.

Now again pronouns evolve around
me as individuals and in coordination,
as multitudes grasp that they are
they, not only he and she,
and not only trans people,
who may be addressed as they want.
All are laminated *they*s,
I embedded, the binary broke;
and with it the disparaging governance
of I/Other : Us/Them crumbles.

You can call yourself your desire:
community scales of performing
assemble us all into they.

This transformation reshapes my perceptions of a unitary self, separate from nature's perfection and beauty.

Suppose non-human nature
is imperfect, improbable, unpredictable
as people. I and our surroundings
assemble similarly in layers, interacting
through our trees of families and resources.

In this construction,
we marks our differences,
alike in certain ways,
different in others.
In this way, *we* rearranges
phenomena, shifting
across social groups to social fabric,
across meals to food groups
to molecules interacting,
to interspecies collaboration
that builds organisms.

Fortunately, difference creates:
Small temperature differences create matter:
linking different molecules creates
life: two different sexes create
persons : multiple races, genders,
classes, and points of view strengthen society.

Similar rhyme create verse:
Similar patterns create style:

Portraits are similar to how we see us:
Differences and similarities:
Identity is multiple
abstractions from events.

Total agreement like complete equality remains entropic and disappears the subject. Even writing at any one level threatens poetries' diversity, societies' readings, and ecosystems' fabrics. Yet within these stanzas, inherent in this use of *we*, lurks a risky *ought* that threatens general consent with autocracy, and myself with losing confidence. But since language continues to be inherently polyvocal, *we* includes *me* and *you*, connected reader. Stress attracts the solo poetic voice.

But can something as simple as questioning what grammarians call agreement and increasing weight on plural pronouns threaten social cohesion? Probably not by itself, although questioning grammar does evoke energetic responses. Rather than fretting over grammatical and ungrammatical constructions, can we accept difference beyond the politics of differently gendered people, showing how these changes in grammar are no less cognitively efficient? According to Julie Foertsch et. al., one does apparently read slower when expected grammatical constructions are not followed.[1]

Selfie asks readers to consider slow reading as a way to understand poetry. For those demanding speed in the cognitive network at every step, consider how reading slower helps support entire classes of people. Consider, too, how rushing often triggers errors. Outside of grammar, questioning inspires reflection *and* threatens fragile connections, encouraging reconnection, multiple connection, and highlights the power of connection. All of these changes are possible and possibly useful in unlearning biases that drive emissions.

I Can Follow Grammar like This
Grammar like this can be followed
by changes in culture and production
quicker than historic rates of social change,
a viral substitution of productive and reproductive process.

In environmental models, the trajectory of agreement differs from the binary world. So does unidirectional progress. *We* suggests that *I* is one of many instances, of many roles people take precipitating individuals and interactions. Not everything, however, is plural and *I* rebuilds trust around probability.

Talking about number calculates
a sustainable understanding
where I and you
are one and many

Here is a foray toward an environmental model. Organisms operate with conditional autonomy to breathe, to eat, to reproduce, and to thrive. As I perform these functions, I separate myself and "sing myself." In this way, beings with outlines, not only through ego, but also through our bodies, initially posit a simple binary: *my* organism and the world, an organismic binary.

Defending my own body against microbes, predators, and intra-species competition facilitates reproduction promoting survival as Darwin points out in *The Origin of Species*.[2] Identifying and defending include such diverse activities as accumulating surplus, building a shelter for security and to secure a mate, farming for food, and sexual selection between individuals. As David Rothenberg points out in *Survival of the Beautiful*, building a bower, making music, and writing poetry are aesthetic functions implicated in sexual selection.[3]

As part of this organismic binary, the mind, through the medial prefrontal cortex of the brain, continually makes decisions about its status regarding that which is and is not self, encouraging conflict, confrontation, and reflection on future consequences of social activities including altruism. Imaging researcher Arnaud D'Argembeau reveals:

> One of the most consistent findings has been that the ventromedial prefrontal cortex (vMPFC) is activated when people contemplate various aspects of themselves and their life, such their traits, experiences, preferences, abilities, and goals.[4]

Here, then, are two ways that individuals are highly incentivized through body and brain to put themselves first and the rest of world after. Depending on perceived threats, others and the world often lag afar.

What does it mean for the biosphere
when the organism is only one
and the world is only zero?
Defensive behavior
explored in psychology
that starts with biology
perceived through social contracts
that evolve into poetry
dependent on habitat support.
I appreciate defended selves daily
at many levels of interaction,
in front of the mirror,
in crossing the street to meet friends,
at poetry readings.
Yet only a suicide would construct
the binary the other way, since
"Being alive means defending a form." (Anne Tardos)[5]

Skin, immunity, nerves, social controls, protective coloration, flight, shade, the list.

Simultaneously, individuals do not thrive alone. Darwin specified three principles of evolution that individuals perform in concert: variation between individuals, transmission of traits, and struggles with environment.[6] My existence, though I perceive it as singular, connects everywhere with many others, all the time, connecting and reconnecting as conditions change. In an essay by Richard L. Summers, the individual organism, society, and habitat continually participate as a manifold of inter- and intra-actions.

> [Humberto] Maturana and [Francisco] Varela note that no living system is ever completely divorced from its environment and knowledge of these conditions is principally required for successful autopoiesis. This knowledge attainment requires an acquisition of information and translation to system meaning in a Bayesian process of inference.[7]

Inferring and defining the self scales between the organism and its reproductive processes by transmitting characteristics to the next generation. Defending the organism scales down to defending ideas and scales up to group defense of territory and altruistic group behavior. While not all processes scale freely up and down, individual organisms operate through self-defining principles of organization, social behavior, and engaging with surroundings.

Defining the self cannot be viewed as purely personal or social because they are built into the biology of organisms by function and form rather than their causes. Sometimes the functions of organisms and their forms like wings occur by chance, sometimes for survival, and often for multiple reasons. Hair and feathers, for examples, both warm the individual and attract sexual partners. Their functions can be defined through their connections within individual bodies, to other organisms for competition and mutual benefit, and to non-biological processes like rainfall and temperature in the ecosystems they inhabit.

Writing through the organismic binary and social/ecological networks makes all writing multidirectional. Multidirectional does not mean reading forward, backward, and vertically, but writing and reading forward in several directions, as me, as us, and as ecological networks linked with all life forms. Rousseau understood this nature as a social conflict for citizens in their dual role making and obeying law, but the environmental model extends these multiple selves to the very action of being alive in the biosphere.

Fortunately, my mind provides me with the tools to realize and shape interaction with others: According to David Premack and Guy Woodruff, humans, chimpanzees, and probably other species have a theory of mind, that is, the ability to infer the thoughts of others that helps define our plural.[8] Theory of mind develops in people from age five onwards. Concentrating on others permits understanding how and why they operate, allowing imitation, learning, and compassion. Concentrating on others helps me understand myself by contrast, a meditation similar to looking at my own mind. Imagine the evolutionary advantage from millions of long nights in dim firelight staring into each other's eyes.

Contributions from other organisms—
cousins, carrots, and gut bacteria
air and water—
reveal a continually shifting reliance
through sets of connections
to the world outside the mind.

Because of this indelible dependence on others, we form groups with other individuals through:

- similarity (poets)
- cooperation (humans & dogs)
- mutualism (humans & gut bacteria)
- commensalism (cattle & egrets)
- parasitism (humans & mosquitoes).

Accepting affinity for others is a successful strategy for individuals. Behaviors that shut down these connections to others and surroundings result in isolation, a dependency on reflection, and a loss of control of habitats as in the cases of mirrors, screens, and climate change. Plurals persist here, too: no one social interaction channel suffices for all situations. In theory, the more connections we make, the more resilient our brains, social networks, and ecosystems. Like organisms, an increased number of connections help individual websites gain more visibility. As Albert-Laszlo Barabasi says in *The New Science of Networks*. "90 percent of all documents, have ten or fewer links pointing to them, while a few, … are referenced by close to a million other pages."[9] Can I apply the word *stardom* to documents? Can I apply the word organism to environment?

Organisms appear in *Selfie*'s frame as an increased density of connections—more neurons in my brain entice me to think of myself. Sometimes I suffer from social anxieties and a weak theory of mind, "and no birds sing." At those times, I don't feel comfortable understanding the motivations of others; I prefer the consolations of reflection to my enduring dread of others. Rejecting even offered connections to others precipitates substantial mental anguish, fragility that may threaten my race, gender, age, and class, and I lose certain compensating skills. Evelyn Reilly:

this ambiguous animal
skittish with the notion of identity
as a series of equal signs
forming any kind of viable ladder[10]

Identifying myself
as only equal to myself
and allowing others to tell me
who I am, who am I? Both alone
create anxiety, while approximation,

the tilde ~ and ≈ (approximately equal to),
are more accurate symbols
of environmental, social, and contemplative
relations than the equals sign = alone.

Modifying Reilly's statement, I'd suggest a series of tilde pairs. A plural almost equal (a manifold with roundabout ways in a congested traffic to formulate approximation) keeps me constantly aware of who I am in places. Plural tildes say I approximate a self in multiple states of being and situations rather than fixed class, hetero, domino, or capability. And organisms tend to age, I being an approximation of who I was.

Self not only exists with others but also extends to groups in several time scales. What I call self, my feelings about it, and its external characteristics are shared:

- in atomic time, too short for humans to register
- in the time it takes for nerves to sense and form words in reply
- in observer time with other selves through desire and lyric
- over biographical time through memory and poetry
- through biological time by inheritance and divergence changing canon
- through historical time building canon since the beginning of records
- through geological time, where poetry speculates, as species come and go and
- through terrestrial and then cosmic time.

Focus on one or more temporal scales changes with conditions:

- the demands of my current body
- impingement of unbeckoned recollection
- associations with family, friends, acquaintances, competitors, enemies, and those unknown around me
- desire to preserve situations, even unpleasant ones
- recognition of transience inducing denial.

These initial extensions of myself—the organismic binary, the need for others that forms groups, and the scalar settings of habitat—operate as collaborative entities, separate and together. My self is not a unit of one and can only be understood as a construction, even comparing reflections on myself of myself to myself, a plural. This shifting state is revealed by meditation on self, social interaction, and experiments in the sciences describing our world.

Categorization arises because each inherent (natural) process has limitations and connections that are personal, social, and environmental. Zoe Liberman et. al.: "Despite the salience of individuals in social thinking, a large body of work suggests that the tendency to conceive of people as belonging to social categories is automatic."[11] Your little dog categorizes you: friend from foe and what to eat.

But all is not perfect in our parsed world. Categories are somewhat consistent within themselves but are also marked by coercion and violence. As Antonio Garcia Gutierrez says in his work on declassification, "The Consensus is part of a dominating effort, which reduces worlds, oppresses perceptions, and maintains hierarchies, putting an end to the dialogue."[12] Environmental models help because observance of plural simultaneity rather than universal consistency undermines human tendencies to dictate outcomes for others. Yet to slow climate change specialists from engineers to poets will change methods while individuals will adapt new consumption patterns. Separation and division of labor collapse only when societies disintegrate.

Connected, iterative selves also need consensus to function and build a world. This paradigmatic conflict threatens civilized consistency, the nation state, and monolithic nature. "The Particles" supports declassification and subsequent reclassification at many levels in operations that install plural viewpoints in how per Erickinson Bezerra: "we relate to knowledge, memory, identity and with the other(s)."[13]

My identities therefore are inherently plural and built from components over time into continuously transforming selves. Component identities—built through multiple groups with internal and external human and non-human partners—also depend on ecological connectivity. There are connections within each organism that allow it to function: blood vessels and synapses. There are connections between organisms that build society: family ties, corporate documents, and political parties. There are connections between organisms, social groups, and the rest of the biosphere that build ecosystems: roads, air routes, tree roots, nitrogen cycle, and language.

Like organisms, poems, including their underlying concepts, do not exist alone, operating in configuration with similar and different other poems for survival, poetry as a class of writing, evaluation, and other network activities in poetic ecosystems. The poem is part of a network needing at least a writer, an implement of construction, a "page," and a reader to exist and to be widely understood as a poem. Oh, yes, and probably a publisher and distributor. If it's paper, the supply chain from the forest comes into play. If it's on a screen, then the whole network of digital technology attaches to the poem as in Flarf (an avant garde poetry group associated with early internet poetry derived through search terms and email threads)[14] and other electronic poetry. If a poem exists only in your mind, then the networks of the brain are engaged in conception, construction, and storage. Later perhaps vocalic apparatus builds an audience that becomes part of an oral network. Asif A. Ghazanfar and Drew Rendall:

> Studies of numerous species have shown that the vocalizations that are used in these social interactions often contain cues to caller identity and that listeners are sensitive to these cues. Many of these same social dimensions underpin routine human social interactions and are cued by the same voice features.[15]

To delay extensive lists of connections for a few chapters, I can isolate a poem conceptually, but poetry is never truly separate from its processes that includes reading and its means of production. My poem, as David Bromige connotes in *My Poetry*, is part of my conceptual framework attaching to other concepts and other frameworks.[16] The poem does not exist alone. The details of poem construction, its operation, and identity under various conditions need elucidation for the environmental model of self and poetry to be useful. Farms do not produce food without friendly climate. Poets do not write without prior poems, including a supportive climate of ideas extending far beyond poems. You, connected reader, are here with me, you poets, you thinkers, you amateurs in our friendly climate of poetry readings, poetry books, and poetry discussion that help create a contemporary poetry identity while climate change looms ever more present in the weather and on our pages.

Described environmentally, the poet and other identities form to help one operate independently and to link to others—one's individual, one's groups, one biological layer 30 km deep ringing one planet with countless relationships. In multiple roles, I act both as an individual organism and as a connector to others, assembled across linguistic, social, and ecological

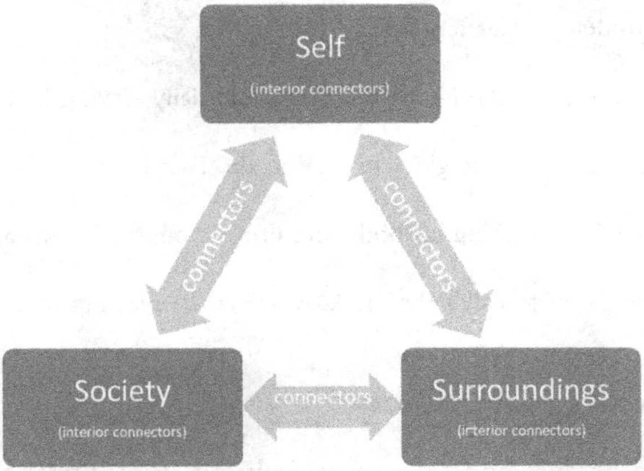

Fig. 1 Environmental model of self, society & surroundings

structures. Here's a start on an environmental model that compares to the humanist model from the Introduction (Fig. 1).

This four-part network entity—organisms, social groups, biosphere, and their connectors—sketches the environmental model of identity. And, of course, this singular shape has multiple instances of condition, time, and location.

The environmental model above adds the term and process of connectivity to the Venn diagram of self, society, and nature in the humanist model. The environmental model draws attention to how people are linked together within selves, to each other, and with their surroundings. Connectors include specific behavior composed of configurations made of palpable stuff and material language, constructions somehow simultaneously psychological, social, biological, and ecological. Linguistic connections are vital to writers who ignore their network and "wander lonely as a cloud" at their peril. In a world where climate changes quickly, including these linking features in the model supports life.

While conjunctions can all be read as doing similar work, linking components increases the value of the connecting particles *and*, *for*, and *so* at the expense of *or*, *nor*, *but*, and *yet*. As Christopher Smart writes to connect himself and us to his pet, he repeats the conjunction "for" as because and proceeding.

For I will consider my Cat Jeoffry.

For he is the servant of the Living God duly and daily serving him.

For at the first glance of the glory of God in the East he worships in his way.

For this is done by wreathing his body seven times round with elegant quickness.

For then he leaps up to catch the musk, which is the blessing of God upon his prayer.

For he rolls upon prank to work it in.[17]

For, as in because and since, even from his institutional confinement and instantiated through and for, as in intended for, his cat, Smart aspires not only for inclusion in society and the biological film around the planet, but also to accentuate the materiality and changeability of language by repetition. To the extent that we perceive individual identity leading in writing and thinking, the conjunctions *or* and *but* dominate, referring to separation from the group. In "An Essay on Man," Alexander Pope writes, "He asks no angel's wing, no seraph's fire; / But thinks,"[18] where the word "but" separates humanity from heaven.

In the following fragment from the poem, "Middle Passage," about loss and disconnection, poet Robert Hayden uses the word "that" as a conjunction until phrases are linked by "with" in death. The feverish link of Jesus with pilot uses space as a conjunction, the subject understood and the verb form of "pilot":

'That said black men could not be reached;
that the Crew abandoned ship, leaving
the shrieking negresses behind:
'that the Captain perished drunken with the wenches:
that two hundred thousand dollars in
prime slaves were lost,
that six white men did perish drunken with the wenches.
'Further Deponent sayeth not….'
Pilot Oh Pilot Me[19]

Today, in an America divisively separated by ideologies fomented to cement control by oligarchic forces, we have a personal and political grammar dominated by oppositional conjunctions—*or* and *nor*, *but* and *yet*—reinforced by our fetishized delusions about separate and independent individuals, misrepresented as free. This fiction had a purpose historically to free individuals from God, from autocracy, from social and economic servitude, but now the pendulum has swung quite far the other way and these same fictions appear as mechanisms of control in consumer society along with fictions about the purity and perfection of "nature," another mechanism of control. Anastasia Thanukos noted how Darwin pointed out that imperfection of the fossil record was a sign of evolution;[20] to paleontologist Stephen Jay Gould[21] modern evolutionary theory has shown that perfection in nature can no longer be assumed. Evolutionist Telmo Pievani of the University of Padua has even written a theory of imperfection,[22] perhaps following Yogi Berra's theory, "If the world were perfect, it wouldn't be."[23]

Environmental grammar emphasizes *and* and *so* and *for*, implying that an individual and their groups operate similarly and together in relationship. But I would not to exclude *but* because we still need to be individuals to operate successfully, semi-autonomously, when necessary, the organism being one of the indelible structures of biological existence. Charles Bernstein: "Poetry and citizenship are inconsolably incommensurable, conjoined at the heart but beating time to different drummers."[24] Language writing as usual limits its range to individual and political frames, but the link "at the heart" is easily extended to habitat, "inconsolably incommensurable" but linked. By linking poetry and politics, language writing builds connective links and the notion of connectivity that this text extends to habitat and ecosystem. To become successful agents of change, the organismic binary and connections with society and habitat may be viewed loosely together as the environmental self within the manifold of self-society-ecosystem-connectivity above.

While the manifold churns out identity, both cultural innovators and sustainers shape it to local conditions. (Here local means within the configurations of multiple related entities, not just a geography.) Some leaders shape language to control, some add healing and sustaining clauses, but all use mutual aid in linked locations sharing similar identities.

ONES for Example

You, Yous
You out there!
Youse and

Me, yes, *me*. Me
Are we and us
Or was.

Twos
And I am we, too,
A this and a that.
Those a they
And thus, an I, they
That is an I.

Threes
Not what you think,
For you are and aren't,
And so together
We are not what
We think we are
We don't see
Ourselves as
We are, plural.
We and you and
I are they!

Fours
What's more they
Is that feared them
And they and I, too,
Tie into multidirectional
Characters of connecting
To a person playing we, and you
Aren't only one rather the
Many too as much and more…

Environmental identity is formed through the connections built within and between individual biology, societies, and surroundings, an ecology. I recognize how, with increasing diversity, this manifold fashions us all as *they*. I suggest that all individuals are plural and frame how particles of language change with plural identity. Things do not occur by themselves, yet the organism remains a key structure of biology. I introduce the organismic binary as a structure that protects the individual and performs identity while at the same time multiplying it through social engagement and interaction with surroundings through internal and external connectors that are many and varied. Events in memory rewrite all four and continually reconfigure their networks. Hence, this summary somewhat alters your reading of "The Particles" chapter that sought general consent in society through lack of agreement in grammar, lack of agreement that acknowledges difference while recognizing similarity.

Nouns & Things: Changing Their Climate

This chapter adds nouns to the socialized particles inhabiting the first chapter. Things in themselves are revealed as componentized constructions modified by our perception of how they operate socially and ecologically. Physics, feminism, Daoism, ecology, trees, and poems all assemble through similar methods showing how they do not operate alone even as people discern them separately. Classification, a strong tendency for people and probably other organisms, both obscures and reveals the connections among things. Ontology and epistemology begin to come together. The chapter expresses concern about Stein's noun-centered (organismic) orientation that attempted to limit poetry to a "vocabulary entirely based on the noun"[1] and ends with a discussion of social ecology models from Jason Moore and Bruno Latour.

* * *

We nouns—
both royal and composite
as far as we're separate—
adapt to the surroundings
of our assumptive selves,
non-selves, and connections.
As Karen Barad points out: "A performative understanding of discursive practices challenges the representationalist belief in the power of words to represent preexisting things."[2]

J. Sherry, *Selfie*, https://doi.org/10.1007/978-981-19-4870-1_3

Twisting syntax to make sense and invoke
other sense, non-sense, living nouns twist social
relations to secure their future,
to empower, to inform.
To say that people don't change
as their habitats change,
to say our surroundings don't change
from industry actions doesn't follow the facts.
Yet sequences embrace more than facts:
desire and survival, perceived and real,
all classes alter their habitats to improve
their chances—viruses, ants, elephants, people.
Changes from climate may be sticky
or passing when applied
to organisms, reform when applied to society,
and volatile when applied to an ecosystem.
As climate changes, we will be transformed
if we don't adapt.

Since there are more nouns than any other part of speech,[3] *conditionality are us. Wouldn't it be helpful to our species if we transformed ourselves rather than be forced to change in unpredictable ways? Our preference for freedom of speech and action so suggests. Our preference for strong leadership, for passivity in certain realms, also suggests our penchant to be socialized. The interacting borderlands of freedom and security are complex compared to either condition. People want both. Conflicts arise.*

Our bodies, themselves things with outlines, and our minds, where outlines are less distinct, tend to organize the world by things. Even conceptual nouns with constantly shifting outlines like thoughts, feelings, states, events, and processes have a thing-like character in language as we identify and use them.

Non-humans also tend to organize their world around things. In speaking about the octopus, Peter Godfrey-Smith finds similarities and differences in "…the ways that smart animals handle the stuff of their world. They carve it up into objects that can be re-identified despite ongoing changes in how those objects present themselves."[4]

The environmental model identifies things from multiple perspectives operating simultaneously and sequentially:

- Physics defines particles by position and velocity.
- Metaphysics discusses things as having outlines and substance.
- Environmentally, organisms are readily perceived as unitary and as parts of ecosystems.
- Linnaeus' taxonomy of species organizes species by morphology, dragging into each category many divergences.

Yet things don't exist by themselves in any persistent or, to use the environmental term, sustainable identity as Immanuel Kant's notion of "things in themselves" points out:

> *Though we cannot know these objects as things in themselves, we must yet be in a position at least to think them as things in themselves; otherwise, we should be landed in the absurd conclusion that there can be appearance without anything that appears.*[5]

Not only do thing-presentations not occur alone, they are linked in complex ways to word-presentations. The organization of the topography of human mental apparatus proposed by Michael S. A. Graziano in *Consciousness and the Social Brain*[6] accounts for the emergence of linguistic structures as things that humans use in speaking and writing. More on this later.

Words and things are able to be identified as each other even though a word's contents are not the thing but from different perspectives an idea of it in the mind, a method used to communicate or present a thing, and its interactions that do not necessarily include something physical to point to. Similar and different; at many levels we see the related and divergent characteristics of things presented as themselves and words.

In order to accurately identify the complex properties of things and, by extension, people, words, and concepts, all these different nouns make more sense from the environmental point of view as composite and multidimensional. They are perceived as units in order to be manipulated, but are assembled from other things using specific connectors. Even things with obvious outlines are composite and constructed by their own and observer processes. Here are some examples why things don't exist in themselves but rather are compounds and parts of larger and smaller apparent wholes:

1. Subatomic particles seem like basic things, and they all have different names. They also have common properties. Yet, in order to be said to exist in a strict physical sense, an observer must know both their distinct position and their velocity. In order to know position, the observer must use a fixed measuring device, and in order to know velocity, the observer must use a moving measuring device. According to Karen Barad's description of the Copenhagen interpretation of quantum mechanics (Niels Bohr's indeterminacy principle, as contrasted with Werner Heisenberg's uncertainty), since these two observations cannot be made simultaneously, the existence of the particle by itself is indeterminate. The particle itself is thereby contingent and composite. The particle's existence needs the two measuring devices to be considered in combination with the particle. Then, the researchers who operate the measuring devices attach themselves to the thing. The particle's environmental proof of existence depends on the observer, multiple other components, and their entire set of connections.[7] (Indeterminacy does not mean that the thing (the particle, in this case) doesn't exist for the greatest trick of the ruling classes around the world has been convincing people that the world does not exist.)
2. I feel like a thing, a person-thing; I am used like a thing by other persons and things. Michael Pollan points out in *Botany of Desire* that humans may also be used by plants, like apples and potatoes, that make themselves important and tasty to us so we will increase their populations.[8] I exist and function through operations between parts of myself, the networked components of my body that allow me to function as a thing, and related constructions that I don't consider me. Connected components outside my body allow me to breathe, write, and get paid by my publisher. I had parents who did things to make me, a doctor who dragged me out, a food chain with farm workers and markets that fed me. This thing—me—acts like a network and an entity.
3. Things' indeterminacy also derives from their different uses. An oak tree is a composite of its own parts and parts shared with other plants like soil and water sources. The oak is also composed of other organisms—nitrogen fixing bacteria, fungi, and viruses—that can be more

or less independent of the tree. A tree cannot be said to exist alone as it breathes, is fed by, and feeds in relation to many other entities. The shared components and processes of arboreal and rhizomic plants complicate the binary of Gilles Deleuze opposing rhizomic and arboreal. Further, the tree exists in different ways depending on which combinations the reader views. The tree has a different identity for and shares identities with:

a. a bird nesting in it,
b. a fungus that supports mutualism; Weile Chen: "predictable relationships between below-ground traits and nutrient acquisition emerge only when both roots and mycorrhizal fungi are considered together, …"[9]
c. a poet that sits in the shade of the tree to write a poem such as Mary Robinson's:

 Oft do I seek thy shade dear with'ring tree,
 Sad emblem of my OWN disast'rous state.[10]

 than it does for itself as a noun/entity/subject position.

4. Living matter, "organisms taken as a whole," expands across the globe through its components rather than as an entity. Yet the whole of the biosphere operates as an entity, an extension of life. Vladimir Vernadsky: "The diffusion of life is a sign of internal energy—of the chemical work life performs—and is analogous to the diffusion of a gas. It is caused, not by gravity, but by the separate energetic movements of its component particles."[11]
5. Poems, too, are conglomerate things composed of letters, numbers, words, punctuation, and space that field poetry reads as duration among other things. Poems exist in minds, on pages, in air, and on screens. The poem-concept includes poetry writ large, poems themselves, poetry processes, and poem parts and methods. The poem on the page exists by virtue of paper and on screen through machine language. Poems have purpose, process, existence, and a taxonomy. When you read a poem, it links to many other things such as:

 a. A prior set of poems that allows the poem its form, themes, various subordinate meanings, and companionship in a book which may be different in a single author book than an anthology with a theme.

b. A set of components—reading apparatus, psychological and other mental interactions—that encourages the reader to read the poem, such as love, boredom, school assignments, anxiety, and hierarchies of value that the reader anticipates and invokes as reasons for reading the poem.
c. Correspondences: the reader and poet connect with logic, memory, immediate sensation, and mediated impression. Both paths of communication and active performance of components operate ecologically.
d. General, received theories about poetry that many readers focus on when reading popular poems by Dickinson, Li Bai, Hafiz, and others.

These components dynamically engage the noun/thing (poem, tree, particle). Noun-things are active phenomena linking to others in processes that manifest moment to moment through pixilated perception. Barad:

> *'Matter' does not refer to an inherent, fixed property of abstract, independently existing objects; rather, 'matter' refers to phenomena in their ongoing materialization.*[12]

The thing materializes from the processes of the network. Barad substitutes "phenomenon" for "thing" which makes sense in the way things' components increase and decrease in number and scope based on how far we map the network. Supported by Kant and Ludwig Wittgenstein's formulations around language, Barad's idea sustains multiple views of nouns, things, and subject positions through grammaring, mattering, and helping to perceive things with wider perspective and accuracy.

> Just as there are no words with determinate meanings lying in wait as so many candidates for an appropriate representational moment, neither are there things with determinate boundaries and properties whirling aimlessly in the void, bereft of agency, historicity, or meaning, which are only to be bestowed from the outside, as when the agency of Man pronounces the name that attaches to specific beings in the making of word-thing pairs. 'Things' don't preexist; they are agentially enacted and become determinately bounded and propertied within phenomena… words and things are indeterminate.[13]

A similar frame appears to Rae Armantrout in her poem "In Brief":

Let the Mona Lisa stand
for private jokes,

passing thoughts, all
you never knew

about your parents.[14]

Isn't the unique "Mona Lisa" then also many, such that the painting takes the pronoun "they?" They is connected to all the thoughts about Mona Lisa as a person and as a painting. Using Barad's term "intra-action" referring to the "mutual constitution of entangled agencies,"[15] I am telling an environmental way for poetry to look at noun/things, even conceptual nouns, as inter-intra-action. Armantrout is alert to this viewpoint. The private jokes in the poem become part of the Mona Lisa painting that I now cannot expunge when I see the painting.

Again, in her essay "Poetry and Grammar," Gertrude Stein proposes that poetry uses "a vocabulary entirely based on the noun as prose is essentially and determinately and vigorously not based on the noun."[16] While I appreciate Stein's style in arguing, her essentialism throws the reader into an arena defined by things with names like a field guide to wild flowers. Prose and poetry share too much for Stein's distinction to avoid misleading.

Writers after Stein attempt to make her point of view seem *objective* as if ideas *in* things were preferable to ideas *not in* things. The purpose of this objectification was to make poems immediate, but had the ill-effect of supporting the organismic binary and its downstream market practices of objectification and individuation. The leftist poetry group labeled Objectivist included many poets I admire such as Louis Zukofsky, George Oppen, Lorine Niedecker, and Basil Bunting. Charles Olson wanted his movement to be called "objectism" with the laudable goal of "getting rid of the lyrical interference of the individual as ego."[17] William Carlos Williams' dependency on wheelbarrows and chickens also supports capital's effort to objectify and individuate. Many poets continue to focus on nouns and adjectives as in Anne Waldman's refrain, "I am an [adjective] woman."[18] Politically, the thing charms poets even when they rail against capitalism, identifying the thing through synecdoche with materialism.

In the same warlike period of the twentieth century, John Cage and Jackson Mac Low (chance operations), Ashbery (ranges of reference over the horizon of things), Armantrout (environmental juxtaposition and entanglement with physics), and a host of other contemporary writers make a point of avoiding the fact-like essentialism of Stein, Williams, and Olson. Contemporary poetries such as language writing and environmental poetry appreciate the connections between things and language in a more nuanced way through a range of other disciplines like philosophy, linguistics, anthropology, biology, and earth science.

Environmentally, the poetic thing operates as a combining entity even as it retains its outlines. The wheelbarrow performs in Williams' "The Red Wheelbarrow" with rain to shine, was my childhood vehicle of choice, and is now a repository of top soil in a drought. The rain conducts, irrigates, and reflects, while the idea resolves, organizes, fixes, and flows through nouns, pronouns, verbs, modifiers, and conjunctions.

Therefore things, partisan,
partial, things, of which it is
written so much depends on,
are one a-part of the other,
ideas mostly between,
not in them.

Importantly, things in certain constructions stand in the way of wider acceptance of environmental models:

- the organismic binary, a primary source of essentialism
- bodily binaries like eyes and hands
- how binaries are already incorporated into some scientific theory, social critique, and religious ideas, making it difficult to re-educate leadership
- the power structures such as political parties and institutions built to defend binaries and their oligarchic forms for good and ill and
- the one-dimensional divisions that imitate property boundaries, that supposedly represent and exclude what we're not.

Just because I have outlines
doesn't mean I'm not parts
and parts of parts

connected
by different energies
from the strong force
to the outline of my thoughts
of you, of our species
and

The environmental construction connecting language, individuals, society, and habitats depends on a set of combining entities—conjunctions, phrases, networks, ideas, beliefs, social connectors, institutions, national identities, rocks, water—that form a complex network. Many writers call this a web, an attractive metaphor that seems "natural." But there isn't a single spider sitting in the center of the biosphere waiting for the fly. What a bad joke and neurotic fixation it would be if our entire planet was modeled on an ego and a victim who might be the same one. I think it's safe to say that most other uses of the word "web" are modeled on this spider trap.

Yet in many ways humanist cultures (Platonism, Daoism, and other mostly binary cosmological simplifications) support the web metaphor. We see similar constructions in misuses of Darwin as Social Darwinism, of sociopathology, of compassionate, benevolent, and dictatorial autocratic politics, of obsessions with beauty, smartness, bravery, and other top-down models of control. People tend to value these dualisms the more their security appears threatened, the current Covid-19 plague and climate change among them.

The spider's web, when looked at in more detail, while still simple and lovely in shape, is more complex than and conflicts with the metaphor. Spider webs have two types of connectors between intersections—non-sticky and sticky—one for spider paths, one for capturing insects—an attractive metaphor but not a very pretty politics. The web catches unwary readers with a lovely "natural" image, metaphor at its most dangerous, by which I mean an inferential trap. Frankly, its charm is lost on me. Although I have always loved Charlotte's gentle rule, I cannot forget her occasional outbursts of anger. (Early readers of this text flocked to the defense of this web metaphor because they have enjoyed and adhered to some ecocritics who use it such as Jason Moore[19] and Timothy Morton[20] who have many important things to say, but the web metaphor demoralizes).

In the environmental model that I'm proposing for collaboration, connectors between things operate with different and similar characteristics like the two thread types of the autocratic web each with many internal complexities. Connectors can be identified through mathematics, physics, biology, society, psychology, and poetry. Connectors can be differentiated by race, gender, ability, and class. They can be real and imagined. In the environmental network, multiple connectors with different characteristics—regularity, direction and strength of flow—attach within and between each noun. Some are spider webs which have an isomorphic similarity (a similarity of form, shape, or structure and the quality of being similar) to the more complete model since web metaphors scale up, down, and horizontally. Metaphoric and spider webs are built by the same forces as the ecological network of the biosphere as well as the unique skills of certain arachnids.

While Moore sticks to the web metaphor, he strangely attacks the value of communications as a web activity, ignoring or denying how both physical networks and network metaphors scale. Communication functions through the web when the spider notices the movements of the captured fly. To communications we can also add travel, capture, and dwelling through Moore's *oikeios*[21]. The risks of autocracy in Moore's idea of "belonging to the household" also appeared in fascism and communism. *Oikeios* risks too much for the environmental model to accept the term as informing all operational uses. It's another metaphor that operates at one scale. The household and household metaphors do not scale, because the similarity among participants and authority that keeps the household together neither supports individual liberty, preserves diverse levels of authority in society, nor adapts readily to environmental change. The planet is not a household of diverse species if for no other reason than the planet has no head of the household. The planetary network characteristically shifts power to where it is needed. The household typically does not.

The entire environmental project risks terminating both democracy and liberty if viewed and operated through oikeios, *limited to household and family, their restrictions and hierarchies.* D*iverse populations with diverse needs and different positions in the power structure are not allowed to fully express their differences and are most often forced either to conform or lose the power that comes from participation. More specifically neither poetry nor political freedom will thrive under a single, dominant authoritative model. I do not suggest that a sense of belonging should be dissipated since similarity like difference and language is a fundamental building block of desire. Instead, belonging does not scale fully either up or down.*

Rather the group formations described below are likely limited to similarity of appearance, structure, and connectivity. The components of social groups and thing-groups are rarely similar, being differentiated precisely because the group needs different skills to function and different materials in building the group's inlands and borderlands. Construction teams are composed of carpenters, plumbers, electricians, and a manager while the construction market contains contractors and those needing buildings. Constructions and assemblages of the biosphere vary in scalability. Some of the most useful constructions scale further through grouping and connection.

Moore shows how "humanity-in-nature" and nature-in-humanity "interpenetrates at every level" into an operational function when humanity is "bundled with the rest of nature."[22] The metaphor, however, becomes murky. What are these bundles? Moore's idea of "relations" seems vague beyond the mere fact that relations exist. How relations operate remains opaque. Ontological interpenetration looks more like the humanist Venn diagram in the Introduction.

Distinguishing Moore's argument about Karl Marx from the environmental model does not make his perspective wrong about the effects of capital on ecosystems. But he focuses excessively on states of being such as "interpenetration" and "nature-in-humanity" rather than the operational methods needed to engage climate change even though he wants to make his book operational. Nevertheless, Moore's analysis helps me understand social ecology.

Beyond the threads and intersections of webs, environmental networks also include nodes at many intersections of their connectors. The nodes are, you know, you and me and others of our ilk and other ilks, things and concepts with and without outlines and, if looked at in a certain way, substance or material. They, and here I could say we, interact at intersections mixing flows through channels from other nodes. These intersections threaten, promise, and sometimes deliver. Together, nodes and connectors establish a trophic network connecting organisms through chemical, generally non-linear, interactions. Nodes, like persons, also act to route traffic, channeling through a social model since energies move freely across them and are translated through them. Other frameworks include both ecological and semiotic approaches that circumscribe Linnaeus' tree model and Deleuze's rhizomes.

Respiration and nutrition are exchanges of energy and matter. In "Song of Myself," Whitman describes a recirculating network, and a tree does that, too, connecting with both air and earth to thrive. Multidirectional writing begins at the beginning, singing both a song of myselfie and choral mutuality in the same verse:

I celebrate myself, and sing myself,
And what I assume you shall assume,
For every atom belonging to me as good belongs to you.[23]

On the other side of the oversimplified web metaphor, Latour and Steve Woolgar's Actor-Network Theory (ANT) addresses the network model in a more thorough way. Since their 1986 publication *Laboratory Life*, ANT has been taken up by interdisciplinary researchers in many fields. ANT postulates that things such as groups of people in social theory don't exist in the way we speak of them. Latour's ANT posits nature and society as "two *collectors* that were invented together largely for polemical reasons, in the seventeenth century."[24] ANT discovers two frames for social studies—the "sociology of the social," encompassing traditional social theory and the "sociology of associations" dealing with both the divisions and connectivity that are central to my present text.

Latour and Woolgar determine that "'society', far from being the context 'in which' everything is framed, should rather be construed as one of the many connecting elements circulating inside tiny conduits."[25] *This formulation has so much in common with my current text that I must rely throughout on Latour as presaging my environmental model.*

ANT's thingness deals with problems in social theory even as it develops something akin (sic) to the environmental model. Latour:

> If technoscience may be described as being so powerful and yet so small, so concentrated and so dilute, it means *it has the characteristics* of a network The word network indicates that resources are concentrated in a few places—the knots and the nodes—which are connected with one another—the links and the mesh: these connections transform the scattered resources into a net that may seem to extend everywhere.[26]

The words I have italicized, it has the characteristics, *imply network is a metaphor. Contrarily, the environmental model maintains that these characteristics also build a physical, scalable network with metaphoric components, extending to show that metaphors are physical. The chapter "Networks of Metaphor" examines this multiple identity of metaphor.*

Meanwhile and again, our perceptions and our organisms are organized to posit nature and society as things. They act as extensions of our organismic binary while they are also components of each other. Moore: nature operates in humans and humans function in nature.[27] The constituent parts of our surroundings and organisms also function without our knowledge. To say they don't exist locks Latour and Moore into binary theory. They are not only abstractions. They exist because they—social groups, social movements, social criticism—can be described by their operations, although their outlines are fuzzy. More about these complex outlines in the "Borderlands" chapter.

Constructing a model of the relationships between selves, groups, and surroundings includes at the outset a collective composed of plural selves, plural socials, and plural ecosystems, both material and abstract. And by extension, per Barad: "the concepts used to describe phenomena are not ideations but specific material arrangements."[28]

Latour comprehends the path of describing our world through networks is a slippery slope.

> ANT aims at accounting for the very essence of societies and natures. It does not wish to add social networks to social theory, but to rebuild social theory out of network.[29]

If the ANT idea is based on essence and alterity, then the thing which Latour wants to define isn't. Instead of continually replacing one thing with another (you reject me to assert yourself), it makes sense to see things/theories/people as inherently collaborative and able to be identified with one another in different ways as populations and uses of them rise and fall. Different environmental networks must write multidirectionally toward the thing/noun/self as perceived, toward our groups of which there are several, and toward the environmental processes built by the combination of networks/conglomerates/components.

Can people operate in these multiple, simultaneous frames? Sometimes!

To clarify these entities, the environmental model needs these provisional formulations of nouns that deal with individual/network identities, social/ANT interactions, surroundings/habitat inter-intra-actions, and connections to both surroundings and selves, an ecological approach. In a practical, linguistic politics, David Harvey illustrates "the incredible political diversity to which environmental-ecological opinion is prone."[30] He cites the following political frameworks that use environmental language to support their theories: authoritarianism, corporate and state managerialism, pluralistic liberalism, conservativism, moral community, ecosocialism, ecofeminism, and decentralized communitarianism. The diversity of poetries practices the same linguistic diversity. The environmental model attaches to many prior constructions and allows it to operate for a while into the future. Senses function, organisms operate, and sentences interact through multiple channels. Channel intersections usually include entities that expose the expected conflicts of allegiance that people in our noun capacity all feel between our protected identities and needs for others. Intersections also conduct the surprising associations that help us make poetry.

While nouns are not
conditional in English,
pronouns like
whoever are.

Verb Solutions Adapt

This chapter introduces the role of verbs, especially the verb *adapt* as a key function of the poetics of climate change. The chapter discusses verbs associated with climate change and what verbs do. Collective, built, and multiple agencies can no longer be exclusively lodged with the independent subject; thus, environmental agency is assembled and distributed. Agency extends beyond selves, groups, and ecosystems to their interactions. Agency and one's actual self-interest are connected by probability. Outcomes in large systems are more predictable than in small systems with their quirky configurations and high-impact outliers. Poetry expands its correspondences by showing ranges of outcomes rather than the singular emotions of Romantic and Anthropocene poetics. Even predictive phrases raise questions pointing to ranges of meanings.

* * *

We humans continuously *adapt* to our surroundings. The terms and conditions of the biosphere *demand* our attention. I even *go out* with an umbrella when it*'s* not raining. Our future as a species *will be determined* by what we *do* to and with our surroundings in the next few decades *to avoid* habitats becoming excessively harsh. These actions *will initially operate* as bidirectional connections to affected ecosystems, but if corporate and governmental leadership *listen*, multiple feedback from those ecosystems *will help to determine* the next steps.

J. Sherry, *Selfie*, https://doi.org/10.1007/978-981-19-4870-1_4

One vital way for poets *to adapt is to begin to change* from primarily *representing* our surroundings in poetry *to writing* with the biosphere, *aware of using* its methods through multiple strategies that *encourage* individuals and groups *to adjust* behavior and viewpoint. Publishers *can present* books with different purposes. Readers *can adapt* their expectations of what they *are reading* and why. I *know* it *can happen*; I'*ve seen* it three times in my life: NY School, Language Writing, and Ecopoetry.

Such adaptive connections between action and language operate as Godfrey-Smith says in two directions: "to link perception with action … and creating actions themselves."[1] The link that connects what we see and what we do is verbs. The action of the sentence also happens to be verbs. Verbs *identify* the things humans *do* to themselves, each other, and their surroundings. From the point of view of the individual organism, verbs *act* in multiple ways. They:

- *alter* surroundings, society, and selves through their actions
- *effect* and *affect* our surroundings, that is, the world *forms* our actions as much as we *form* the world for better and worse through a manifold of environmental influences…
- *combine* with other entities *to assemble* composite forms by *linking* and *expending* energy: what we *say* and *do* to each other, how groups *are formed* by affiliations and similarities as well differences like bi-parental reproduction and corporate division of labor
- *hold together* any combining entity to *help* it *coordinate* its internal activities such as our bodily functions, our thoughts about ourselves in a Cartesian way, quantum action at a distance, and within and through thought and sensitivity to sound and reference as in writing poetry. What we *perceive* as ourselves….

To act, I *construct* and am *thrust into* multiple roles, unaccustomed positions. I *perform* variously. I *travel* existing channels *to act with* worlds of people and non-people. My right hand *moves* toward yours in a channel that *starts* from my left sole.

How many things can a noun/phenomenon *do*? *Can* the woman *blink*? *Can* the cat *crochet*? *Can* the cat *cradle*? *Can* the signing gorilla *think*? *Can* the politician *tell the truth*? *Can* anyone *tell* the whole truth and nothing but the truth? *Are* verbs also *subject* to conditions like nouns, adjectives, adverbs, and phrases? What *acts* on climate *change* besides writing?

Verbs *attract*
locations and conditions by *combining*
in syntax, a multiple, constructed
agency of verbs and things where
culturally sticky phrases like
the Anthropocene, the noun, *perform*,
where non-humans *assemble*
with humans to pollute, where
people *adapt* to survive,
connected to food crops
and animals, bacteria and viruses,
air and water, freedom and responsibility,
composed of multiple
connections, within and without,
to how organisms *perform*,
how we *imagine* and
wonder where they *perform*.

Human actions—create, pollute, expand—
threaten non-humans shaped and
shared by those verbs—
reproduce, feed, shelter, survive—
which connect to the Anthropocene
The term adapt refers to vaccines
that act when culture adapts
so harmful effects on habitat galvanize
organized resistance to human authority
like a pandemic. Verbs
adapt through tense, number, and
probabilistic condition.
Adapt, the verb,
combines with a subject,
also with time: "adapt now!"
Subjects imply no rigid states
but inter-intra-action.
Apple, too, adapts.

When results I expect do not occur, I become querulous and worry less what I do to get results, resulting in pollution, thoughtlessly effused. I adapt by resisting, developing alternatives. I invent things with words that I find in my mind, between us, that my worldview presents to me, and those that words merge into hybrids. I write sentences to articulate action. Sentences frame themselves, one word after another, building narratives and referring to endless variety and profusion of meaning. Darwin calls this condition "an entangled bank."

Following the tangles at the text level. Verbs bridge. Following further, verbs link nouns to define a thing's existence—velocity and position. The bird nests to define the identity of the tree for the bird. Here the verb to nest connects the bird and the tree. A nest is also a noun that precipitates several actions endemic to it. The tree might feed the bird with its berries and protect the fledging from a raptor. Every noun participates in an agent combining with others, an ecology.

Verbs generate nouns in state: "the tree falls" produces a falling tree. In this way adjectives arrive. Without verbs things remain inchoate or as Bohr says indeterminate. Verbs propose how things come into being for each other with specific practices. Verbs determine.

Then one action such as turning my head changes what I sense next. We act, as Godfrey-Smith says, "to control what our senses will encounter."[2] Survival: the multidirectional connectors between verbs and nouns are all required for things to thrive. Yet what seems to be a healthy practice such as fertilizing a field to feed hundreds looks different when I turn to see the runoff that causes the pond to stagnate. Verbs do different things in different places, their value determined by connections and their location in the network.

You readers, you poets, you thinkers, you doers, and I all know now that we must act differently to stop the planet from turning hostile to human habitation. We cannot continue as before. Do we want to stop the planet from turning or nurturing? Can we adapt? Appropriate actions are difficult to come by with many political and agential reasons not to risk acting.

1. Research shows there are typically five reasons people don't want to act on or even talk about climate change:

    a. I don't know enough
    b. I don't want to talk about scary things

c. I don't think I can make a difference
d. I don't want to cause an argument
e. I haven't got time to do more than survive.[3]

2. The ecology world, the poetry world, and popular culture are awash with different conceptions about the perception of, nature of, effects of, and interactions between social and individual behavior as well as points of view about the value of inaction, stasis, meditation, and attentiveness. Williams' "The Poem as a Field of Action" contrasts with Auden's "Poetry makes nothing happen." Revolutionary poetry and Buddhist poetry appear in the same anthologies. Language writers and slam poets appear on the same stages.
3. Computer and internet feedback support our organismic binary's misconception that cyberspace is effectively reality, supporting human bias for immediate response rather than planning. To post on social media seems to some their action whereas posting produces no result beyond taking people into corporate electronic space, increasing the prevalence of autocracy in governance as people opt out of the decision processes that constitutional citizenship guarantees.
4. We make mistakes. Restrictive environmental regulations may cause vested interests to react to equalize perceived and target levels of risk reducing our ability to change.[4] This risk doesn't mean we don't need to regulate, but instead to develop awareness of how change impacts ecosystems. Humanity does not operate perfectly and neither does the biosphere. They operate imperfectly and in ranges that are often wider than may be comfortable. Corporate leaders are particularly risk averse. Once they have established repeatable cash flow, they are loathe to try something that might inhibit profit.
5. Corporations and wealthy individuals have figured out how to suppress action against capital accumulation by, as Moore suggests, denying a "conceptualization of capitalism as already a relation of humans with the rest of nature."[5]
6. Humans praise aggressive action on behalf of our independent agency, freedoms, and rights. The politics of resistance asserts the right of civil disobedience. We protest and assert our equality in the political arena, in the workplace, in our homes, and in poetry. We assert our rights to kill other animals and plants for food, cut

> trees for shelter and warmth, and dig the earth for farming and mining. Can we adapt to make these acts less destructive to our surroundings? Can poetry adapt to share the limited space on the page, on the stage, in the classroom for multiple styles? Can poets stop understanding poetry as only expression of self and society?

Considering all the difficulties that acting on climate change presents, we must adapt. That's the verb, to adapt. I adapt, you adapt, they adapt, we will one way and the other inevitably. (Although there are some who think that human resolve/agency/projection is the problem, I don't agree because all life has agency through its organisms and acts to survive. Action is life's syntax, the verb to combine).

* * *

Acting to slow climate change depends on who will act. Most individuals barely affect climate. Corporate leaders need a rationale to act and resist change as much as I. The general concept of agency proliferates in an environmental model from individual agent to multiple agents to composite agency through verbs. I am a component of an agent of transportation when driving with my car, but when I'm sitting here typing I am an agent of writing. Agents build composites and inter-intra-act within an evolving taxonomy, evolving with the manifold, containing currents contrary to the flow.

Matter, especially living matter, isn't passive and performs in my identities. Things emerge through specific inter-intra-acting as with the bird and the tree, the poet and the poem, the corporation and its supply chain. Agency multiplies even as the organismic binary focuses attention on the heroic actor, the poet, the genius worker, the great nation, and the revolutionary concept.

Classification creates problems for agency. To extend the idea of things emerge through what Bohr/Barad call "apparatuses," agents are not merely in the world, they are made of world materials. Refracted through the discipline of particle physics, Barad calls agents,

> material configurations or reconfigurings of the world that re(con)figure spatiality and temporality as well as (the traditional notion of) dynamics (i.e., they do not exist as static structures, nor do they merely unfold or evolve in space and time).[6]

Further complicating agency, organisms operate effectively and independently through discourse even as they are entangled indissolubly on other scales. To address the question of who acts on climate change, agency becomes more than the ontological choice between Barad's Agential Realism, Deleuze and Félix Guattari's Assemblage Theory, and Latour's Actor-Network Theory. These are not alternatives, but environmental perspectives from particle physics in Barad, ontology in Deleuze and Guattari, and political ecology in Latour. When I engage for a specific task, when I change my manufacturing process, when I write a poem, I emerges from the process of writing as a poet. I am not until then. Later, I emerge as a citizen when I vote by engaging myselfie with Latour's network, emerging as myselfie via Barad, and emerging in my territory via Deleuze and Guattari and more recently Manuel DeLanda. Finally, I emerge in the manifold as a bounded and variously connected organism.

Simply put, the environmental model questions the extent to which a named individual, group, or ecosystem can act as an independent agent. The model does not predict their state of being. States change by reconnecting with environmental conditions that are actions themselves, that emerge from the manifold of self, society, and surroundings through connectivity.

If the self is constructed by and emerges from multiple components of individuals, social conditions, and habitat traveling established channels, then the environmental subject likely operates with a different, combinatory agency when performing the unaccustomed acts required to address climate changing. Agency operates as a product of change, of specific actions that are grammatically incorrect but have a recognizable syntax, to be and to seem, linking verbs as action. I reduce attention focused on diagnosing individual agency, because selves construct agency in parts with others using verbs, the connector transforms into an action. Physically, I'm becoming as I inter-intra-act. Organisms socialize with the network in one or more specific syntaxes. Ecologically, connecting is the act of being alive.

Emerging as organisms inter-intra-act, the constructed subject with its complex borders and properties adapts and executes other actions to divert climate from its increasingly unpredictable performance. Verbs act to fashion a specific subject like a poet, worker, and owner of the means of production who now perform in multiple roles in various networks. These subjects extend through their connections as apparatuses, assemblages, composites, and components of the network. Verbs act with nouns in emergent processes. While I continue language more or less as it is, I let you know that the complexities of environment have changed how we are seen/seem to act in situ. My syntax is comprehensible but sometimes altered to accommodate reconfigured relationships and intersections. More substantial changes in syntax and diction appear in language writing such as Peter Inman's reconstructions:

Opus 30
Thinking stein upside down
Someone where the applesauce ~~put it~~
Whatnots to a landscape's hair[7]

Inman's shifting discourse allows me to adapt to conditions. I can't help it no matter how I resist. My identity adapts to shifting roles as I progress during the day, year, lifetime. Since performance enacts from multiple points of view, I increase my chance of understanding what to do as conditions change.

Subjects perform from geological layers, built slowly depending on language deposited in contexts and their availability at the moment, fragile and fugitive, emergent and safe, especially when facing machines rising and planetary systems crumbling while we watch individuals and groups hardening their borders against change, hardening against each other. Poetry that performs ecologically can use complex and often problematic diction and language structures, language technologies, that make reading intentionally more difficult breaching habitual limits that writing remains about communication. Poetry restricted to oneself and self-expressive language enables social fragmentation and ecological insults.

At a little lower layer, all sentences are already in some sense self-referential. Myself is layered phenomena constructed from multiple points of view of the connected apparatus. As with multidirectional writing, self points inward and outward to non-self in both social and environmental ways. Multiplicity are us! Poetry will always in many ways participate in that pleonasm as in Shelley's "One Word Is Too Often Profaned" or lines by John Milton that T.S. Eliot felt were too extended:

Thrones, Dominations, Princedoms, Virtues, Powers,
If these magnific Titles yet remain
Not merely titular, since by Decree
Another now hath to himself ingross't
All Power, and us eclipst under the name

Of King anointed, for whom all this haste
Of midnight march, and hurried meeting here,
This only to consult how we may best
With what may be devis'd of honors new
Receive him coming to receive from us
Knee-tribute yet unpaid, prostration vile,
Too much to one, but double how endur'd,
To one and to his image now proclaim'd?[8]

At other times, poetry's tropism for circumlocution, intended to align diction and syntax throughout a poem with the complexity of events, is balanced by the minimalist aesthetics of haiku and Pound's plea for "DICHTEN = CONDENSARE." A concise example from Lorine Niedecker in "Poets Work" both uses and addresses condensation:

Grandfather
advised me:
learn a trade
I learned
to sit at desk
and condense
No layoff
from this
condensery[9]

Modernist succinctness in poetry, non-emotional geometry in Bauhaus architecture, and other minimalisms in visual arts mirror the need for speed and efficiency in capitalist industry and autocratic efficiency. How can poetry today extend beyond the self and society to the biosphere without including concision and extension in the right context? *Adapt!*

* * *

In "Poetry and Grammar," Stein says that prose focuses on verbs and adverbs. "It is wonderful the number of mistakes a verb can make and that is equally true of its adverb."[10] And I might add true of many adjectives. The pressures of global war affected Stein's perspectives and choices and similar pressures are building today. We see the mistakes; we rush to correct them. We make other mistakes and build them into paragraphs: "Sentences are not emotional but paragraphs are."[11] Through the emotional spaces between sentences in parataxis and even through the emotions generated by linking multiple complete thoughts in hypotactic writing, emotions are one of our animal responses to our surroundings, in this case reading. National emotions must be acknowledged as a social force in World War II Germany as well as twenty-first-century autocrats and wannabes, even if emotion is not a rigorous category. How can we not lament strong emotions from climate change? Who will praise these disruptive changes?

When something goes wrong, many seek someone to blame. One solution for the oppressed of the world might be rebalancing of power, but disruptions like revolution can easily result in more oppression, more powerful autocrats, and support for those restrictions from those who just want to live in peace. Norman Fischer's critical, satiric lines mimicking those who would readily accept tyranny. "Let me be—or let there be—lets all be—/Crowned with radiance".[12]

If prose as Stein says "can be the essential balance that is made inside something that combines the sentence and the paragraph," then the environmental model's presentation of prose and poetry at the same level renders that something prose is inside of a language ecosystem. And that suits me. "Prose is bound to be made up more of verbs adverbs prepositions prepositional clauses and conjunctions than nouns."[13] But as pointed out in the prior chapter, while the majority of words are nouns, the majority of words in prose and poetry are not nouns. Thus, no one part of speech supports Stein's evaluative hierarchy even as she builds a part-of-speech-based system that scales up to an ecosystem. Too often, then, even as a Modernist Stein reverts to her self-assertions in essential, totalizing ways that weaken the environmental position that her thorough approach to parts of speech in "Poetry and Grammar" implies. Romantic assumptions about the universality of art further obscures what might be an inclusive path.

Fighting for environmental justice within corporate and government organizations often means speaking to them in the emotional terms that Stein suggests through paragraphs and I'd add stanzas. Environmentalists are unlikely to succeed at driving investment to green solutions in global corporations without adopting some of the functions of language used by financial leadership as a vehicle to make decisions, resolve disputes, enact practices, measure results, as well as share innovation and creativity. That is, poets cannot expect much environmental traction without some measures of instrumental use and intention. Fortunately, poetry can decide as in Robert Frost's "The Road Not Taken." *The Iliad* satirizes failures of leadership decisions and extols the virtues of Odysseus' rationality.

Actions taken by humans on themselves, actions taken on others, and actions not taken by humans that affect them must adapt to be understood as probabilistic rather than only binary, analogic, and representational channels. Poems are built around such ambiguous and indeterminant language and structure through Keats' negative capability. In another discipline, in "Improving Inductive Inference," Richard Nisbett et. al. predict: "We envisage a program in which probabilistic models and statistical heuristics will eventually be incorporated into most people's everyday reasoning."[14] The facts move toward poetry.

Nisbett and his coauthors attempt to show how adaptive culture might be able to move us toward probabilistic and rangy solutions to the problems confronting us: "One reason for optimism is that human reasoning changes with new cultural inventions.... The modern notion of probability is scarcely more than 300 years old."[15] Nisbett et al. suggest something similar to Bohr/Barad's idea of emergent phenomena. People:

> do not retrieve attitudes from storage at all. Rather they construct them from materials at hand, including their current affective reactions to the object, their semantic associations to the object, and observations of their recent behavior toward the object.[16]

This ad hoc approach to thinking entangles the corporate center, those primary contributors to and remediators of climate change where process modification must be implemented. Corporate leadership continues to defend both science and rationality. Scientific consensus does not appear to be widely disputed among corporate leaders, although there are some outliers in the fossil fuel and chemical industries.

But two kinds of rationality—rational self-interest and environmental rationality—confuse the discussion. Rational self-interest, where the disruptions of climate change drive leadership to protect assets by avoiding scientifically sound climate policies, conflicts with environmental rationality where, as David Sloan Wilson opines, "Adaptation is the gold standard against which rationality must be judged."[17] For example, many corporate leaders vote and pay for tax policies aligned with their short-term self-interest, rationality in the narrowest sense, while at the same time they support converting to electric transport, wind and solar, sustainable farming, and environmentally sound mining and construction. Diverging corporate interests appears as the dominant conflict in today's political economy.

In addition, while corporations dialectically contain the problem and solution, they are losing a battle with non-rational (in both senses) internet speech that now dominates daily opinion for many citizens in the US and worldwide. The cash flows of the supposedly liberal Silicon Valley leadership support such non-rational modes of discourse. In these multiple, self-defeating ways, rationality and pollution are aligned. Narrow self-interest of individuals and expediency, rather than environmental rationality, justify the cultural reasons why we as a civilization are not leaping to defend our ecosystems against immanent destruction by insurgent pollution.

Economically, the same conflict rages. Rationality is associated with self-interest through representations of human behavior in neo-classical economics writers like Adam Smith and Milton Friedman. Fortunately, contemporary behavioral economics points out that rational self-interest is infected by conscience, linking individual allegiances within groups, multi-level selection, and recognition of the value of ecosystems to self-interest. Works by Dan Ariely,[18] Herbert Simon and his notion of bounded rationality, Richard Thaler and Cass Sunstein,[19] Daniel Kahneman,[20] and other writers show how self-interest is multi-level.

The connected reader realizes that multi-level self-interest, like evolutionary selection, reiterates those actions of ecology that are isomorphic and scale like connectivity from sub-atomic particles to the biosphere. Self-interest is not about being lonely and making money as Ayn Rand's Objectivism would suggest. Self-interest is several components of a combining phenomenon known here as self that adapts identity with each role and iteration of role. The phenomenon formed by combining self-interest with other components may not, however, be at the scale of a species like homo sapiens but an apparatus like an organism, a machine, supply chain (combining organism and machine), a philosophy, or a poem.

Environmentalism projects and absorbs how poets can adapt to reduce risk and still progress their languages through the ranges of scale that poetry facilitates with methods like metaphor, synecdoche, and metonymy. Writing that soothes by avoiding elements of linguistic surprise attracts many readers. Nevertheless, plans that assume climate and language continue as they are today protect my view of myself while they fail to protect me. You see, they, ourselves, and our acts are not the same but similar. Adapting requires attention to probable results, using verbs in different ways. Adapting requires adopting the likelihood that complex configurations like poems, selves, organisms, populations, and ecosystems produce events that can only be predicted probabilistically in ranges and rarely as single values. The benefits of adapting exceed the energy to understand experiments in language. New ways generally become tradition as pointed out in *The Climate Report*.

> Integrating climate considerations into existing organization and sectoral policies and practices provides adaptation benefits. Further reduction of the risk from climate change can be achieved by new approaches that create conditions for altering regulatory and policy environments, cultural and community resources, economic and financial systems, technology applications, and ecosystems.[21]

People adapt to conditions to reduce exposure, reduce sensitivity, and increase adaptive capacity.[22] Integrating our surroundings and the rangy ways, we view and use them into both Frank O'Hara's personist poetry ("vend astringency off hours,")[23] and the sociocentric poetry of twentieth-century experimentation (Bob Perelman: "Naturally enough I turn to./Some things are reversible, some./You don't have that choice./I'm going to Jo's for")[24] incorporates habitat by challenging linguistic norms. But there are many ways to catch a fish. Not that poetry is everything, but it adapts at different scales like other processes in the biosphere. Poetry continues to change. We can and do talk about poetry with other processes operating at their own scale attending to conditional differences—poetry and war, poetry and love, poetry and ecology.

For poetry to support mitigation of global warming means adapting to new conditions of language and adapting the language and syntax of poetry to new conditions in society and your local ecosystems. We can preserve the past under glass in received language, forms, and modes of address, but an ecological culture encourages adapting rather than stagnating as in *The Climate Report*: "Interactions between adaptation and mitigation strategies can result in benefits and adverse consequences."[25] Complementarity of culture and active engagement are critical to human survival.

Damages and the opportunity for poetry to address cultural changes to labor, extreme temperature mortality, coastal property inundation, air quality, washed-out roads, intermittent electricity supply and demand, inland flooding, failed urban drainage, rail disruption, toxic water quality, disappearing coral reefs, waves of new illness, dying freshwater fish, end of winter recreation, collapsed bridges, harmful algal blooms, Alaska infrastructure swallowed in mud, dying shellfish, crop failure, increased aeroallergens, and wildfire all remain potential to poetry, not only how I feel about it. List absorbed into text from *The Climate Report*.[26]

Specific structural changes to the page appear in Jennifer Scappettone's *From Dame Quickly* and her extensive work on copper and climate change, Jack Collom's *Exchanges of Earth and Sky* on birds and people, Reilly's *Styrofoam*, Joy Harjo's "How to Write a Poem in a Time of War," Juliana Spahr's *That Winter the Wolf Came* addressing the Deepwater Horizon and Occupy, and works like Pico's *Nature Poem* and others are just the tip of the iceberg of possibility.

Marcella Durand sent me a list of forms that she thinks can address climate change structurally including: Aleatory Poems (Tristan Tzara), Cut-Up Reassembled, Eclogue, Ecstatic Poem (Kit Robinson), Elegy (Michael Gottlieb's "The Dust," Haiku, Insult Poem, List Poem (chapter "Borderlands"), Matrix Poems (some of Mac Low's *Light Poems*), Non-Ambulatory Walking Poem, Reshape and Update any poem, Reengineering Prepositions ("The Particles" above), Sound Poem, and improvisations and journals (Kit Robinson or Alan Davies). The field poem, developed through Stéphane Mallarmé, Pound, Zukofsky, Olson, and more recently language writers from Grenier, Robinson, and Bernstein to Myung Mi Kim and Caroline Bergvall, encourages ways of reading that aren't simply left to right scanning. They allow readers to read up, skip, and scan diagonally across the page to discover a renewable and changeable syntax rather than a fixed path or goal that was critiqued as early as Francesco Petrarch's "Ascent of Mont Ventoux" in the fourteenth century. But Petrarch's forays across the field have been short circuited for hundreds of years by power and the requirements of polemic. Now I write another path that you may follow but the path is multiple, like the roads to Mecca, like the probabilities of weather forecasting, like the ways the self operates in society and biosphere.

The traditions of individual identity alone and identity of/with others informed by regular old sense data persist as evolutionarily advantageous myths that are necessarily unfaithful to the richness of the biosphere, because phenomena like people operate through their connections to other organisms in situ and in action as simple as turning our heads rather than only as things/people in themselves. Are we able to simultaneously act with both the myth of individual identity in order to capture the self and the recognition of nearly universal interaction in the biosphere in continuously emerging phenomena? Neither one can function effectively alone.

Now we can see
from one thing to another
how adapting methods
of connecting can become,
well, natural, by which I mean
inherent in all the performances
in the biosphere. How we adapt
to global warming appears
itself in parts of speech like verbs.

Borderlands: Am I I & They? Knowing Without Or

This chapter thickens and broadens the boundaries between and outlines of individuals, groups, and their surroundings. The environmental model of borderlands specifies a multi-dimensional set of frontiers composed of multiple connections. In contrast with the one-dimensional property model, these borderlands contain specific channels between people, groups, non-humans, places, concepts, and the processes of the biosphere, connections both ecologically interactive and isomorphic. This chapter's multi-disciplinary—psychological, political, aesthetic, scientific, and textual—list of boundary conditions supports multiple interactions at multiple scales through different disciplines operating between components of the biosphere.[1]

* * *

> The U.S.-Mexican border *es una herida abierta* [an open wound] where the Third World grates against the first and bleeds. And before a scab forms, it hemorrhages again, the lifeblood of two worlds merging to form a third country—a border culture[2]

In *Borderlands/La Frontera* Gloria Anzaldúa details the suffering of mestizos on the US/Mexican border, "a dividing line, a narrow

J. Sherry, *Selfie*, https://doi.org/10.1007/978-981-19-4870-1_5

strip along a steep edge." Underneath the tragic stories on the one-dimensional border, she writes about the busy complexities, the beauties, diversity, otherworldliness, and multiplicity of the ways people relate to each other in these wide and dangerous borderlands. I am pleased to adopt her use of borderlands as a broad, fertile area of variety, cultural change, and responsive interaction through her many identities—female, Chicana, queer—in prose and poetry, multiple identities through multiple languages—Spanish, English, and Nahuatl, adding the specifics of ecological connection to Anzaldúa's stories and descriptions.

Standing in the way of implementing environmental models of culture that support change and creative difference, inland humanism and the laws of nations depend on reinforcing the separation of me from everyone else, separating us from them, and establishing opposition between humanity and nature. Like nations, Catholic believers endlessly discuss good versus bad. Daoism presents a yin/yang, black and white world separated by a narrow, wavy line, a separation that is actually wide, eventful, and diverse. Like religions, political models also insist on similar oppositional distinctions. Can poetry compose a more complete model of our world?

What actually transpires across these borderlands between individuals, social groups, places, and concepts? The interior structures of states and religions built to defend prerogatives of class, gender, ability, and race as well as limiting identities to pre-selected political agendas and artistic personae, like The Poet, also impede awareness of ecological connectivity. In order to avoid:

- autocratic misrepresentation of value in social interaction, politics, and culture,
- the closed system of Rand's Objectivism in financial and right-wing political circles,
- adherence to points of view that treat exclusion as necessary to their beliefs,
- narcissistic dysfunction among poets, and
- simplistic views of boundaries represented as only what we're not, a line separating me from what is not me,

expanding our perception of borderlands becomes increasingly imperative. This "Borderlands" chapter shows how, in some denoting and comparative detail, individuals, groups, places, and ideas extend into each other as part of the multidirectional flows within, to, and from organisms, groups, and places in writing.

Borders are broader,
edges are entries,
margins marginalize and connect with page numbers,
limits limited range and shorter paths strengthen connections,
property lines align and conform.
Let the sets presently reset assumptions
how nouns and verbs proliferate
without telling us what to do with them.
How to engage more specific speech
as inhibition approaching borderlands?
How may I address you?

In order to free writing to explore what has been excluded, blocked, and erased by extensive in-country descriptions, borderland zones adapt to geography through their components more than property lines. Borderlands absorb difference. They diffuse and diffract cultural thrusts, redirecting similarities toward differences that can be absorbed approaching the other side. Borderland zones connect one thing to another through overlapping, dense arrangements of composites such as we. Edges adapt, interact, and even create since variation is there most apparent, since we organisms see through difference.

Frame borders as things with width,
depth, height, and extension over time,
the border's borders even broader.
What's going on in
that space between
that can ally us?
Prose becomes poetry
then reverts to it:

Borderlands
swell from one-dimensional lines
through the compound model of
things including molecules,
nouns, individuals, nations, concepts,
ecosystems. When we say
our organism exists, can we say
so for all parts or only when they
come together in the known

identity individuals construct
of things and selves? Scale
falters in perception
persisting. Borders expand
through thinking about them.
A person can be said
to exist, even described in parts,
and may be conscious. But is
my arm conscious? An octopus
may be said to be conscious, and,
per Smith, there is a greater likelihood
of extending consciousness
to its arms through its distributed
nervous system, distributed network, populations.[3]
My brain is dark
inside my skull, but connections
fill me with light.

Graziano: "The truth is, we all have phantom
limbs superimposed on our real ones."[4]
The brain controls attention.
Selective signal enhancement
controls the body.
I turn my head to hear better
but your face remains where
it was in spite of me moving.
Try it.
Existence in the mind
extends
the body's existence.
My gender reassigned:
I seem to be there
like
Lord Nelson's arm,
an illusory presence![5]
Prose becomes poetry as we,
as we traverse the borderlands..

The metaphor and physicality of boundary protect and separate unforgivingly, No Trespassing signs nailed every few trees. The self-absorbed and sociocentric writings of Romantic and Modernist poetry bring linear borders up close and personal to arouse the suppressed emotions protecting selves as property. Through institutions such as marriage, war, free labor, and slavery, people function as chattel as well as sensible beings. Robert Frost: *Something there is that doesn't love a wall.*[6]

Extending Frost's conflict about property, Susan Howe's notion of boundary *refracts* multiple characterizations of borderlands through questionable history. Howe refines the ways one individual is separated from another. Alienation (dis)connects people from themselves, their group(s), their nation(s), and the biosphere. A clear, bright border of identity appears, different for each person in each condition that protects the organismic boundary while fuzzy interpenetration usually remains the actual case. Howe implicates writing in this boundary:

MARK / border / bulwark, an object set up to indicate a boundary[7]

There's no good reason to think of sharp divisions except to facilitate more precise operations and to protect you and yours. The COVID-19 virus makes this dual purpose apparent to me as I stay home to protect my health and the health of others. With one gesture, putting on a mask, I speak for myself and for others. When I protect myself, however, I expose myself to tyranny, allowing my groups to advance unrehearsed into my territory. The risk of losing myself in myself in this situation remains the odds-on favorite.

That, too, is multidirectional writing.

Corona Shadow
As buds break into new leaf
Cloud shadows sweep the hills
Their moment of darkness threatens
Nothing will stop our spring

The region where two or more entities connect is hardly a one-dimensional line. The open wound of Anzaldua's *Frontera* gapes as wide as human suffering but the healed cut is strong. Laminate and evolving as identity, variable and sequential as syntax, networked and simultaneous as situation, borderlands compose lives. Boundaries are places themselves and orient our attention to the actions performed in them. Consider how the bundles of channels that network individuals with their surroundings, themselves composed of connections of different types, traverse the membranes of self, self-image, body, behaviors, and thoughts of the surrounding ecosystems, of other organisms, other composite things, locations, and other, even jarring, ways of thinking.

I know you now, but later you will be a stranger.
I love you forever except when you annoy me.
Please come and work for me
until I make a mistake and fire you.
until I make a mistake and fire you.
Trade with my country until you gain
an advantage. Then we can make a war to
profit from our people's suffering.
I need you when I'm hungry.
Surely you know these stories.

The soil allows us to attach ourselves;
the world allows the delusion of detachment.
Attachment allows us to escape the illusion
of outside; detachment allows us
the escape of borders. Self presents
the balancing act to be refined.[8]
I reflected and now like
my reflection perhaps too much.
I store it in the back of my mind
and it comes forward untoward.
Across the street I saw her inky noodle
shattered against the vitrine of reflection.
Seeing one thing through another:
refraction and reflection both
operate across borderlands. Put a pencil
in a glass of water. See the giraffe.
The space to the mirror shifts as I tilt my head.

> *Methinks that in looking at things spiritual, we are too much like oysters observing the sun through the water, and thinking that thick water the thinnest of air.*[9] (*Herman Melville*)

At another scale in another measure, borders become an essential component of humanist identity, certainly national identity. Open borders, superficially a good alternative to border walls, relinquish control of governance to unresponsive and self-serving corporations whose own borders, the corporate veil, are highly defended and far more difficult to define and breach.

> *When the power of the state is taken away, only commerce, its corporate prerogatives and battles with labor survive. Crossing borders becomes a financial transaction, porous to capital, facilitating, according to David Harvey, "the free flow of capital across the surface of the globe."*[10]

At the same time, border crossings become restricted to the value of labor to capital. The risks to human agency of corporate-only governance argue for multi-lateral trade and labor agreements where multiple interests—individual, class, corporate sector, and state—are negotiated by those organizations that remain responsive to citizens. States mediate differences between citizens. Even in these days where the state seems beholden only to corporate interests, labor and individuals still have a seat at the table. To some extent the state supports labor's interests in negotiation with employers. Thus, the state remains a required intermediary for citizens. Harvey:

> *The fierce class and traditional resistances to capitalist modernization in Europe, on the other hand, made the intellectual and aesthetic movements of modernism much more important as a cutting edge of social change, giving to the avantgarde a political and social role broadly denied them in the United States until after 1945.*[11]

In specifying this problem of borderlands, isomorphism continues to be a valuable tool to understand the complications of borderlands. The isolation of the body of the individual is mirrored in political identities instituted in the seventeenth century. In the Peace of Westphalia:

> *…old forms of hierarchical deference were quietly discarded. The inherent equality of sovereign states, regardless of their power or domestic system, was instituted."*[12] *H Kissinger, writing on the Peace of Westphalia that established the notion of equal national sovereignty, misrepresents almost all of the primary issues involved. About the same time, René Descartes was writing "cogito ergo sum" and the East India Companies, chartered in Holland and England, were receiving rights to isolate their business from direct political control and competition. The border remains a wound scabbed over by walls of nations, ghettos, and unreconstructed disciplines of thought.*

Even linear property lines and national borders are more than one-dimensional creations, established through legal arrangements, topological features, individual expectations, and cultural differences. Most nations support the right to own private property, yet laws vary within nations.

Further intricacies arise when property is threatened: the fear, the rage; owners protect property as if it were their own body. They imagine themselves defined by a line; don't cross it, noli me tangere, argh!

These diverse situations regarding interfaces between things—cell walls, house walls, disciplinary boundaries, physical solitude, and ecosystem boundaries—are not easy to resolve with singular restrictions. Cultural theories and most ecology texts cited here attest to the materiality of the network of complex borderlands that surround and allow independent identification of atoms, genes, cells, bodies, thoughts, communities, institutions, nations, and ecosystems with some scaling up and down. The environmental model of borderlands always contains something we can point to, but it is not a thing in itself any more than any other noun. It is not a one-dimensional dotted line separating you and me. To paraphrase Jacob von Uexküll, each person senses the boundary in two separate but interdependent spaces: the space they effect and the space they see.[13]

Did I forget the middle ground quickly closing between predator and prey, love and hate, transparent and opaque, unit and fragment? Quickly closing! Did I destroy that interim space to preserve what I mistook for myself?

The material and multiple connections that comprise and cross environmental borders at many scales are as multi-dimensional as the vaunted interiors of minds in modernism. Rather than trying to invent a generalization or definition of borderlands, here are some exemplary ways to look at them. Some of these connections are explanatory, some are conceptual, some communicate the border crossing through poetic indirection:

1. Connections across borderlands may be point to point like whispering a poem in your ear (material being moving air), distributed over several channels at once like social media (material being electrical energy across a wire), and broadcast like cellular technology (material being microwaves) and solar radiation (material being the radiation).
2. How amazing that text connects.
   I missed that scaled 与, that fish

who escapt the lure. Twice!
Slipt in votes fer them,
suppressing ballots of Browns.
Yes and so one is
writing in two directions at once.

3. The palpable, measurable border between my hand touching yours. Its heat, touch, run up the arm, focus the eyes, open the connecting capillaries, supporting connections to biology, law, sex, and signs.
4. A taste you suffer
Where th flame ends
Variations tho still one.
Differences linkt thru assonance,
rhyme, those different words, drowning
like Gerard Manley Hopkins ' "six coiffed sisters" in
setting, prosody, logic, and page.
5. The many layers of skin (three major, multiple minor), blood vessels, nerves, hair, fat, integument. I kiss your cheek covered with bacteria. My lips splash through layers of increasing temperature, denser pheromones, resistance by your mind even when you want that kiss on your cheek.
6. The Seine River where sunlight ends
No, look, it sets here, too!
7. The border between El Paso and Ciudad Juarez changes thickness with changes in political administration. People, freight, and ideas (less, then more) continue to flow across the border. Poetic border crossing and overlapping ecological border counter the political border line reinforced by a new wave of white selfhood.
8. Oppression of the Other
pushes strongly in one direction.
Responses to oppression—
connection in Black communities,
assertion of southern lands,
monolithic feminism and queerness,
submission, questioning
dialectic, acknowledging
universal disability, asserted

consciousness, multiplicity—contradict a singular view of self,
society and planet

9. The indirect connections of
a reference made to a friend
about what you did.
To whom. The words were
spoken and difficult
to retract since the words
are material, remembered, archived

10. Between I. A. Richards' "tenor and vehicle," between
George Lakoff's "source and target" connectors carry
thrown light, black dogs, etherized patients, and world
stages between sectors of the brain and words of the text

11. Characters in novels
speaking to each
other on lawns

12. The soil allows us to attach ourselves and our plants
while the world, too much with us, provokes
detachment. Attachments evaporate the illusion of a
great being outside. Detachment helps us escape the
delusion of borderline.[14]

13. "A dark green raincoat she ran into"[15] (Zhao Si)
Pouring a glass of water overwhelms the lip.

14. Fruit flies dance out our door.
I catch you saying that again,
again you must be somewhere to mean it.
A selfie shows what
was behind you and around.

15. The intermittent but regular
connections of taxmen reach,
the toll taker receives,
collection plate passes,
the date waits for you
to take or not to take a pill.
Many to many

16. Connection breaches the personal/social
boundary linking many to many

17. Is there any connection
between a screen loading

and recall? Is there any?
Shape to it!

18. The strong connections of motherhood,
the weaker yet still material connections of fatherhood,
the connections of love,
clearly material and based on fluid
exchange, chattel and use value, the different
connections of friendship supported by various roles
such as long relationships, jobs
and interests, friendships:
enjambment due to other friendships

19. Divine destination in your eyes
Somewhere I, we, they: the heavenly host-
ess cupcakes, hoping not to be toppled
by twinkies.

20. The model links characters in theater, actors on political stages and ghosts that haunt our demonstrable things and those that exist most strongly in language and sensibility, the connections across lines of thought that science proves and those it doesn't validate.

21. How far do I have
to go to get to
where you are
so we can be here?

22. The strong connections of a Black man to the daily pain and pride of his color. The characteristics of that connection—job performance, family stability, class connections—when pain bleeds into self-doubt, what is spoken about as color. The resilience against pain fashioned by community...

23. Driven by dread of distinction
"However the image enters"[16] (Audre Lorde)
"The perversity
of separation, isolation,"[17] (Amiri Baraka)
"An intensely slanted approach..."[18] (Edward Bland)

24. We Jews learned our lesson. Shepherds, slaves, then millennia of separation from Christian societies where my ancestors lived. Trying to connect through borders in Europe, we joined the elite and the working classes.

Separation reasserts itself in holocaust, six million dead
and Jews flee to Palestine, the US, Argentina. Now
Israel reproduces that ghetto wall against attacks from
Palestinians whose name the land bears. The wall
ignores citizenship, familial ties, human rights to protect
individual, threatened, Jewish bodies, corporate
productivity, and the corpse of the state. Reconnect
despite the risks of threats to life and tribe.

25. A baby-boom generation protecting truth at all costs
engenders a generation intent on no harm at all costs.

26. The connections of youth to age, anger to fear,
power to privilege, sex to love that
reiterate binary identity!

27. Beauty connected to arbitrary choices, &
to patterns of calcium propagation, &
to realms of mathematical possibility, &
to economics by our willingness to write
through activation and inhibition signals
through air, through mind, with light, in random
placement of color on her feathers,
as in poetry we inhibit grammar
through connectors to politics,
deleting "the",
to metric apostrophes,
to prosody adding through which
readers awake their aware.

28. I wrote a personal poem through me
a social poem through them
an eco-poem through us
connecting through a fourth line.

29. Bi-directional connections of gender.
A sport teems. How strongly
class distinctions and daily
activities determine and
reestablish connections
to build a social network,
flexible borderlands.

30. Your self is not a unit of one;
Multidirectional writing: that's the whole poem,

because any word could appear next.
It's just a fact of writing
and if you object that a perfect wor(l)d
should appear next or some synonym
then you haven't been paying attention.

31. The intermittent connections to corporate identity
reinforced by marketing, advertising, and public
workers and management while severing connections
between workers that strengthen their communities.
Supply chains imitate family ties and forest ecology, as
well as slave trade routes.[19] (Stefano Harney & Fred
Moten)

32. Corporate identity links consumers and suppliers of
product to the corporate body that acts like an organism,
acts like a city, acts like a mind.[20] (Marsilius of Padua)
But cities are harder to kill.

33. Connections are nouns
made up of components
an over populated dmz
to make want disappearand create desire
like all things connected
and simultaneously separate,
a characteristic of language.

34. "I don't remember."[21] (The Cure)
Obsessive productivity
recapitulates the self.

35. The connections across boundaries between one poem
and another both of which I wrote. The connections and
borderlands between poems I wrote and those you wrote.
The connections and social borderlands between a poet
and a group of poets, casual and formal, connections
across social boundaries between poets and a group of
non-poets, walls between poets and markets, connections
and walls (borderlands) between poetry and society,
connections made here between poetry and the
biosphere, what I'm writing here and speaking now
connecting to you!

36. Multiple uses of the same thing,
pliers to open and close,

tie otherwise diverse functions out there in the field
and inside in the head together:
enjambment as a mental connection

37. Links don't just exist at the edges of entities. Many begin from the core function of an individual's love, family, blood flow, property, and poetry. Interior-to-interior connections are *as* crucial to the organism *than* those beginning at the edges of entities because every organism is composed of many linked parts. More connections within the organism than between organisms help define its identity.

38. One more poem here!
If you write it, then
we're a thing, the
enjambment of sex.

39. These are my personal boundaries
Do not cross them…

40. Some things are connected and function together like a fork and spaghetti. Some appear to be connected by being in a related category, but the connection is not functional like a fork and a banana. Some things might be functionally related but are not usually used at the same time like a claw hammer and a sledge hammer. Some things are functionally unrelated and don't appear together like a kangaroo and a hammer. Some things are functionally unrelated but have other relations like a giraffe and a pencil: remember connections to the glass of water.

41. *Breaking and entering*
into the paradise of me,
suffering in my domain,
reminding the preacher that
life is surplus energy
that founds
&&& causes warming.
Why do I want to sustain
that image, that paradise
of glass that poetry

Borderlands composed this way become porous and permeable, scary to the highly defended self. They are difficult to contemplate as a unit since they are composites of diverse connections, diverse characteristics. Yet they are a commonplace unnoticed.

The poet sees this coming and protects their lyric.

Biology and psychology connect the same organism differently. How do I get to be one then? I am multitudes. They can be brought together, designated together. Epistemology that separates exclusively is not an enemy but understood as linear and, if remaining exclusive, threatens the future of the planet.

Hold your biases close to see me being we.
Cut your connections but continue to look proud.
If you push your biases toward me,
can we be we?
Can I be we,
royally and a composite?
The one is still two (self and other)
plus one (yourself as you see you),
society, habitat, and connections.
The foursome of the environmental model
questions singular and binary identity.

Inversely, the humanist notion of connection defends private property, unique personal identity, linear borders, and accumulation of capital behind those borders. Myung Mi Kim:

> *accumulation of land maintain household bear labor of house child*
>
> *cooking reserve line belonging to elaborate isolation.*[22]

Individuals and groups have political and economic rights to what they own by virtue of expended energy and resources, but not at the expense of the collective and community needs. It's settled law in most jurisdictions and the US Constitution guarantees "just compensation" when property is "taken" by the state. The environmental description of boundary does not necessarily threaten private property, since as Jeffrey Winter points out in his study *Oligarchy,* "free, unfettered" use of property has never really been the case. Winter stresses: "It is important to draw a sharp distinction between a property claim and a property right."[23] But individualism

arising from Romantic resistance to industrialization has developed to the point where oligarchy no longer allows the laws of the state by themselves to protect their property.
Hence Donald Trump, Vladimir Putin, Xi Jinping, Narendra Modi, Recep Tayyip Erdoğan, Abdel Fattah Al-Sisi, Viktor Mihály Orbán, Rodrigo Duterte: I pass their pictures to you.

Poetry of only selves, poetry of only society, poetry of the two, and poetry only about nature support the node-like perspective of capital against multiple, different, quaternary views of borderlands. Poetry of nature, poetry of nature, and individuals support the node-like perspective of humanity against nature. The extension of capitalist metaphors into art, into identity, present a risk to the planet, enabling unquestioned use of resources without acknowledging the interactions that result. Yet it would be difficult to immediately exclude them as they permeate our language use, however careful I have been.
To paraphrase Niklas Luhmann: Environmental borders are semi-permeable as is the rest of the network. Its system is closed but accessible through programming.[24]

And why would selves be any more inviolable under capitalism? Fearful landowners think their selves must be uninfringeable, identifying with an established value like property-like states of mind, property-like ideas like nation states. Should poetry imitate property and the state? (When, as a publisher, do I change my mind and want academics who cite my publications to pay for the right to reprint?) Existentially, the values of property fluctuate. The facts about values vary. History shows little immutable value beyond narrative and now even that is questioned. Where does that leave the defended, immutable self? The real wilderness is the space between us. Marie Buck:

and the best you can do
is cut a sock in half
and try to create something
I'll want to have.[25]

Syntax or How I Become What I Seem

This chapter links the sequence of words and the order of biological development. Identities are also inherited through genetic sequences, learned through cultural legacy, and conditioned by habitat. Selves and social groups build through a sentiency syntax, but minds can progress contrary to the flow, using adaptive thought, and events can be non-linear. Selves progress within known ranges, so looking at other people, events, processes, and languages provides guidance as much as looking at your self. Self progresses seemingly in episodes with memory rewriting the past many times. Writing can flow in multiple directions—to elucidate itself, to progress its themes, to illuminate its author, and to seep into the audience around it—sometimes all at once.

* * *

Syntax herds thought,
assessing self through sequence:
"Am I blue?" I am blue!
"Word order = world order"[1] Steve McCaffery & bpNichol

Different and similar
coordinate recent and disturbing
impacts of some narrative
or weather; you choose which one.

J. Sherry, *Selfie*, https://doi.org/10.1007/978-981-19-4870-1_6

Selves, too, are layered
in biographic sequence,
constructed/construed in episodes,
embedded in linked lines,
phenomena distorted
by frequent diction surprises.

Words in order signify
my reaching, measuring

sequence with social
and environment tools at intervals:

sounds, metrics, senses, sense,
as police say, priors.

Time passes swapping
priorities in the order of sentences

and what-we-say arrangements.
As I shape this

body, a series of glimpses,
rearranges words and limbs,

ties one to the next through prior
phrasing revealing what order predicts:

endangered surroundings!
Selves develop lamellar armor,

punctuated flows into an I,
attaining continuity and permanence,
from difference and distinction.

I don't abide long
in one role, state, or self,
but repeat similars,
building identity approximately:
sex + *gender* ≈ *identity*.

Other identities extend the formula:
race + class + ability +++,
each laminated, constructing myself
by multiple streams merging.

Divisible, too, so one
like seeks another like to group
differences into a similar.
Categories being partial:
they simultaneously are
and are not one, building
society, extending ecology.

The syntax of identities
is commutative with no fixed
sequence of categories. For one person
she (sex) + *they* (gender)
might biographically and relationally build
into an I. Alternatively,
one from many or many from one
might function in reverse order ≈
"the category...permanently moot."[2] (Judith Butler)

I ellipsis'd out the "is" to make it clear that
categories are both useful and dangerous
with provisional and engaging character.
 (both words and symbols
 perform commutative discourse.)

Difference operates between substances (for which there are no opposites, rather another form) and qualities which have both opposites and shading. Such construction is inherently non-binary. A similar condition pertains on the scale of writing:

I can write with almost
any order of words.
But while syntax is freer
in some languages than others,
once started, precedence
increasingly predicts subsequent
order. Narrative,
style, logic, and the writer's desire
to not stop the reader from reading
altogether propel regularity.
Syntax as order is not simply
a state of being as in Warlpiri
where syntax depends on inflexion
in the process of speaking.[3] (David George Nash)
In English, even in texts intentionally
irregular and decentering, syntax
persists to materialize
connection to the reader,
(dis)orients the reader and supports
with predictable comprehension.

Syntactical connectors are poorly defined by art, hard science, and social science that focus on the nodes of organisms, groups, locations, and poems. Nodes operate in less distorted states when included with their connectors. Current efforts by groups focused on color, gender, class, religion, ability, and intention to create conditions of greater social equality selectively discount these links that form specific order. Connections are not allegorical but physical and persistent channels between building syntax.

* * *

In this chapter, syntax operates forward like time, bodily development, and by extension the sequential processes in the biosphere. There are also recurring processes in the biosphere that have both linear development and cyclical development like seasons, political affiliations, and perceptions of oneself. And this chapter focuses on reading and writing that move forward in multiple directions, toward self, non-self (of which there are many), and writing. Bodies move multiply, too.

Linear processes of identity are modified not only by social performance and individual psychology but also by a biological reproductive capacity through bi-parental sex in animals and many other organisms that appears to follow the arrow of time. From this biological perspective, order is significant and determines the majority of cases of development of fetuses. Developmental syntax determines biological being and also the temporary, alternative, and contradictory tendencies that operate within bodies that can be separately distinguished.

For example, Sheri Berenbaum and Adriene M Beltz: Fetuses begin to develop male characteristics starting about the sixth week of fetal development, in some predictive ways even earlier, long before they identified what it means, and even before parental behavior began to dictate their gender.[4] My awareness of gender functioned earlier as a boy child, sexual awareness coming later. (This is only my memory of my experience influenced by parental and other social colonization of my body.) But bodies operate in more complex ways with more flows than I imagined. New information shows how the mother influences gender even before she actually becomes pregnant. Lisa M. Thurston et al. write in a study dependent on both human and non-human species:

> Historically, the sex ratio of offspring in mammals has been regarded as solely dependent on whether an X or Y chromosome-bearing spermatozoon fertilizes the oocyte. Recently, however, data from ingenious experiments showing that X and Y chromosome-bearing spermatozoa elicit differential genomic transcription by oviductal cells have suggested that the female reproductive tract has an active role in selecting which spermatozoa reach the oocyte.[5]

Apparently, a woman's condition, driven by chemistry, by assumptions, and perhaps even desire (although that stretches the facts on the study), interrupts the linear determination of fetal sex/gender. More generally, this study shows how social gendering can disrupt the smooth order of events that determine the sex organs of the fetus, inverting simplistic expectations of developmental syntax. (I bring this up to demonstrate how sex and gender mutually influence each other and then go on to talk about poetry in a similar way.)

As you can see, connected reader, these parentheses introduce my own idea as opposed to my interpretation of the data supplied in the quotation. In this way, perspective and discourse supplement not only the creation of the fetus (that already is identified as a combining entity developing in a specific order) but also the order of writing, writing's assumptions, and perhaps even the desire of this poet to link poetry to the body and by extension to environment.

Contrary and disruptive flows to biology and writing play an important role in slowing climate change. The alternative policies proposed by organizations such as 350.org, NRDC, and Extinction Rebellion parallel ways or reorganizing language flows in poetry. Juxtaposed perspectives can be close like demonstrate/point, but also distant. Non-linear syntax plays an important and frequent role in language writing. Clark Coolidge in *The Maintains*:

mother order
for general past part the feet
each upon which its tone
often a like even[6]

And later even divergent pairs imitate other generative combinations with an increased complexity and distance between words that still generate meaning. Here's Tina Darragh:

Charlie Chaplin charge-a-plate
oatmeal object trouvé
dictaphone different
pidgin piggyback[7]

Pairing takes place first through prioritizing diverse types of words and diverse combinations. The desire of poets and mothers includes their needs for themselves, their poems/progeny, and the characteristics of their affiliations with the potential reader/father as well as the physical level of the poem/zygote which are combined entities that poets/mothers nurture. Complexities abound and predict a wider range of possible discourses than I normally imagine. Because the scale of climate change is difficult to imagine, these complex ways of reading literary and biological creation help understand how far from the norm climate may diverge. Will we be able to read our climate?

* * *

Beyond desire and gender, specific roles often determine the order of my actions.

As a poet order is always
being important, one word
after another. And
time, location, and motion
work-together on social planes
when citizens change polluter's minds
inspiring me to write to continue,
quite unlike the Enlightenment poets:
"His time a moment, and a point his space."[8] (Pope)

I wrote this verse linking syntax to society to self because one-directional connectors, in this case reading forward from word to word, that define important sequences for individuals are often ignored or intentionally concealed by complex framing. Even reading forward, I retain the residue of prior words imposing them on the new word. This process accounts for many of the multidirectional wonders of poetry through alliteration, internal rhyme, and juxtaposition of similar sound as in poets from Hopkins to Edwin Torres in "[cell division]":

compartments

compartments of artment

come party—artment

[*childment, woment, manment*]

many *ins*

going where *outs*

won't—to emerge

redrawn[9]

I read Torres' lines as form, back and forth, up and down to confirm the connections between his words, repeating, breaking, taking my reading toward the shape of his field. Multidirectional syntax points to reading as an individual, as one of the readers who understand Torres' method, and as part of the surroundings of avant-garde literature, categories, and the ecosystems of writing about identity across borders.

Torres is networking multidirectionally
  as an individual and member of several
  language and poetry groups.
Torres is networking multidirectionally
  in writing from himself as an individual poet, as a social poet,
  and as a denizen of his ecopoetic habitats.
Torres is networking multidirectionally
  in the processes of the biosphere,
  that direct his overall themes,
  sequences of events that supply his details,
  and the processes of him, functioning
  better or worse as a writer of investigative poetry.

Now a political example: obscuring connections to donors and negotiations with corporate stakeholders by campaign rhetoric allows greater freedom for leadership from skeptical and importuning citizen voters. Politicians and voters read and repeat the rhetoric but what is owed by the politician to a donor may be more important in determining policy.

Take the following sequence paraphrased from Roots Action: Jared Kushner's real estate firm asked the Qatari government to finance a skyscraper in New York. When the Qataris turned him down, Kushner, as part of the Trump Whitehouse, supported a blockade of Qatar led by Saudi Arabia and United Arab Emirates. The region, already violent and chaotic, was plunged into greater confusion. Kushner then reportedly blocked efforts by Secretary of State Rex Tillerson to ease tensions.[10]

Is this sequence causal?
We don't know, but
the Trump administration
did not want it transparent.
It reveals multidirectional order
public policy builds
from national and financial interest
highlighting oligarchic self-dealing.
Significant political and poetic syntax
is often accomplished in the dark
like careerist sleeping around.
Unrelated sequences can be
promoted as facts
and risky syntax obscured.

My mind, too, conceals and distorts
connections that inhibit
directed action, like measuring
my self-interest by location and direction
as if I were a sub-atomic particle
against industries overwhelming our habitats,
like white is all colors but we see
a colorless translucence.
Distortion scales to misunderstanding.

John Laitner et al.:

> Yet, to date, the majority of climate mitigation assessments have tended to mischaracterize the behaviour of the economic agents [through the deficiencies of neo-classical economic theory] whose decisions ultimately affect the planet's climate.[11]

Psychology tries to expose subconscious
concealment, only superficially addressing,
through a syntax built with the therapist
(which push sometimes does the trick),
the reason I benefits in the short term
by ignoring the herd of indeterminate.
elephants on the mental table:

As in high school I read
"The Kingfishers." Annoyed
at its sentimental "who cares
for their feathers now?" I leapt
to the conclusion field poetry
freed time from meter
to include the value
of time on the space of my page.
Through environmental models I
finds a space for a person moving
with changing acceleration through
time, including, but not relying on, not
denying, any recherche-du-temps
separation of what only exists together.

Here the mind works as part
of the body. Addressing bodily
needs rather than parsing
functions as in this chapter's progress,
seeing it so, to see what's up
with looking at it this other way,
my body in its space connects to others.

In "A Noiseless Patient Spider" Walt Whitman shows how his poetry explores the space around him, reading to discover connections to others: connectivity uses space. (Of course, I still question the web metaphor, but this poem writes the spider's organism in space.)

A noiseless patient spider,
I mark'd where on a little promontory it stood isolated,
Mark'd how to explore the vacant vast surrounding,
It launch'd forth filament, filament, filament, out of itself,
Ever unreeling them, ever tirelessly speeding them.

And you O my soul where you stand,
Surrounded, detached, in measureless oceans of space,
Ceaselessly musing, venturing, throwing, seeking the spheres to connect them,
Till the bridge you will need be form'd, till the ductile anchor hold,
Till the gossamer thread you fling catch somewhere, O my soul[12]

Multidirectional syntax of sentences, individuals, societies, and ecosystems sends bursts of bodily, socially, and habitat-determined identity through their subnets into the networks of the biolayer. The angle of separation from the organism increases as we read while different parts of speech accumulate and jam along those connecting links, across borderlands listed above. Self-identified and other-identified groups connect like the individual connects within, some link automatically, some are intentional, none thrive alone.

* * *

Similar not identical syntax defines us in biology, politics,
psychology, philosophy, mathematics, and language These
multiple syntaxes should come as no surprise.
Similar not identical syntaxes reveal how the biosphere, society,
and individual thought patterns work through specific,
related sequences.
Similar not identical stress occurs in illness, racism, self-hatred,
and just living day to day.
Similar, not identical, syntax reveals what's poetry and what's
prose with stress on word-to-word order.

While continuing to perform these similarities with the complex apparatus required for the execution of poems, we ignore many correspondences and complementarities. As I have said, desire causes climate change. We cannot eliminate desire except by dying, although some religious practices work hard to undercut this primary function of the body in order to reduce suffering. But suffering from climate change does not come very much from the mind. Society and individuals find ways to sublimate or turn desire toward improving the common spaces of society and surroundings. As Norman Fischer might have said although he didn't quite:

Emptiness is sky
Sky is emptiness
Emptiness is not sky
Sky is not empty
All four orders play together[13]

The object Galileo Galilei sees can be
construed as singular image,
but only exists in multiples:
he sees it at many moments,
others see it, many concepts of it.
Can I plural that singular?
Shantideva writes agents operate
together performing their existence:
Phenomena exist other than we see, swapping
identities, transforming through various,
similar roles, yet continue to exist. Engage

in giving for the sake of obtaining
the fruit of your efforts, identifying
yourself as the agent of what, of what?[14]

Getting the order right: things in themselves have lost their sovereign logic.
Getting the order right: the direction of climate change must be reversed.
Getting the order right: we start with our connections to others
    and by social dynamics move in a somewhat determined,
    somewhat purposeful direction from there.
Getting the order right, we do not reason from outliers.

* * *

There are more possible than actual sequences, the plethora of choices a fetish: The DNA code—G, C, A, T—syntax operates similarly. While there are 64 combinations of three letter words, there are only 20 known amino acids. Are the other words valid, too, that is, might they produce life, lively life? Ethical restrictions currently limit research into these words that poetry allows. What is the moral? Can poetry's free exploration help genetics?

Inherently, some combinations
don't work very well
in words, in organisms,
in societies, and ecosystems.
Limits are predicted,
transitional species
erased from the fossil record.[15] (Darwin)

While many particles and nouns can be reordered
for logic, for poetry, some combinations simply
don't / work / well / now.
As much as I am attracted by commutative
personality, society, and property,
they are not fully so.
Order and scale continue to matter.

Order Continues to Matter
He told me only that he liked that poem.
He told only me that he liked that poem.
Only he told me that he liked that poem.
He told me that only he liked that poem.

In addition to processes where some of these possibilities have no physical instance in the world, surprising multiples operate continuously, simultaneously, and often create conflicts for both individuals, groups, and surroundings.

* * *

We desire freedom.
We say we desire freedom,
but ecosystems, with mixed open-closed
formations, are not as free
as we can ideally imagine.

Each step toward ideals
must be a real one.

People today, needing reduced risk,
fearful of change and outside influences,
seek a more determined
(read safer and securer) order,
a syntax that clearly pronounces
the composition of groups
and the relationship between table
of contents and contents.
They assume they will be heard.

We turn, therefore, to English sentences and their sequence even though the period creates deceptive separation, even though English sentence order is more determined than some other languages. Here then are English sentences read through an environmental sequence, a syntax of syntaxes:

They is one person composed of many organisms, roles, and group identities.

While the first pronoun above disagrees with the verb, the particles come together to identify a person in the grammar of a sentence. The pronouns multiply: *they*, *you*, and *we* but not any grammatical default, a preference. I cannot prefer for you, yet relationships remain polyvalent, that is, I am not always being we as well as I.

They is one person composed of many organisms, identities, and roles gagging on complexities and conflicts, often ignoring them for expediency.

One person, one self, ourselves a phenomenon composed of many organisms: 90% of our cells are non-human in a human shape. To paraphrase Scott Gilbert et al.: Biologically, the separate organism isn't unitary or individual.[16] I could stop here, to let you have all these biologies careening through your body, but there's social dimension as well. Within that organism composed of many organisms, many roles also thrive. It takes more than one layer of identity to make a person like me: white, male, aged, US citizen, father, son, husband, poet, editor, publisher, tenant, car owner, registered Democrat, pensioner….

They is one person connecting around even when alone through the bodies.

Taking different roles, I alter my view of myself toward my surroundings, since I think and act differently as a poet than as a father than as a citizen voting. As a poet, I perform with language. As a father, I help perform with language. As a voter, I perform a predetermined task with language. Each of these efforts connects to other writers, critics, publishers, and readers as opposed to Michel de Montaigne's landmark statement: "I study myself more than any other subject. That is my metaphysics, that is my physics."[17] That is humanism.

They is one person; connecting, then, within one and to another are the syntaxes of which we speak.

Connections between individuals, within their parts, across society and from all to their surrounds create a further syntactical arrangement. In the biolayer ecology is the study of connection. In literature the connectors within individuals are lyric poetry. Trish Salah:

> Everyone died, including the dead. Imagine it. Imagine the unendingness of your own dying in and of the interconnected world, all the world of it.
> And to be other deaths in your emergence from the living, because, well time is also a conversion of energy and matter.
> Leave it at that level of clichéd abstraction if you know what is good for you. [18]

Between individuals and their society is epic and narrative, although personal narrative might read as lyric. Virgil:

...thanks to cruel Juno's relentless rage—and many losses
he bore in battle too, before he could found a city,
bring his gods to Latium, source of the Latin race,
the Alban lords and the high walls of Rome.[19]

Between individuals and ideas is metaphysical. John Donne:

Alas, alas, who's injured by my love?
What merchant's ships have my signs drowned?
Who says my tears have overflowed his ground?[20]

Between individuals, societies, and surroundings is the environmentalism of which we speak. Brenda Iijima:

over-systematized then some river
now exoskeleton remains, now behemoth, now erectile dysfunction
now infrastructure, now affect behavior, now collective interest
/now connective tissue, psycho-social-ecological registration
now/methane, carbon, nitrous oxide and ozone...[21]

Viewing individuals as ecosystems within themselves acknowledges how the networks of self—nerves, blood, thoughts, and sensations—depend on sequential flows as well as the network flowed through. Biologically, we simply aren't the individuals we are socially and never have been solely social. The biological basis of these collective identities appears as a product of how bodies are constructed. Widely discussed and always in syntax as being includes becoming.

Sight follows lines,
waves drum hearing,
touch requires, finally, touching.
The most delicate quanta of touch,
felt by the tongue, utter,
triggering organisms' attention.
Attention builds a schema.
Attention and schema interact
as consciousness, to make others visible
as self and other, to organize a landscape.[22] (Graziano)

Being part of a couple that produced offspring was important and restrictive to me, and to my child to which couple he owes many aspects of his existence. As a child he often wanted me to clarify the order of events and what priorities his parents used. Citizens want to know the story, too. Corporate leaders must accommodate the entire syntax of the biosphere in their production processes.

Social Groups Scale: Identities & Connections

In this chapter, the dynamics between individuals and social groups identify ecological correlations. The individual's role guides their thoughts and actions. Sometimes people with similar roles in different groups have similar points of view despite group differences. Poets share poetry even when they write in different modes or diverge by class, race, gender, ability, and age. This chapter encourages readers to include the connections between individuals, groups, and their surroundings as the fourth component of their model of the world. While these connections are material, multiple, and diverse in their characteristics, they are often ignored in favor of the individual organisms that we identify with. I note finally that cultural change often drives climate change.

* * *

Writers often assume that something convincing at one scale operates with many of the same properties at another scale. I want to write like this, even though it's only occasionally the case. So, to start this chapter I take a quotation from James Baldwin's *Nobody Knows My Name* that refers to America as a country and as an idea and transform it, by deleting the beginning of the sentence, into a statement about identity:

> —this collision between one's image of oneself and what one actually is is always very painful and there are two things can do about it, you can meet the collision head-on and try to become what you really are or you can

J. Sherry, *Selfie*, https://doi.org/10.1007/978-981-19-4870-1_7

> retreat and try to remain what you thought you were, which is a fantasy in which you will certainly perish.[1]

Individual identity and social identity have some similarities and some differences. While Baldwin's harsh alternatives make sense for individual psychology through the organismic binary, the environmental model proposes that it is very difficult to be rid of the "fantasy." It continues to creep into my thoughts. When the fantasy appears, the best I can do is identify it and develop skills to investigate what is more likely the case. Despite conflicts arising from this psychological pixilation, I might also call it inner commensalism, that accepts and practices continual negotiation among multiple interests increases my ability to adapt to new conditions.

Environmentalism pursues not only this solution to handling difference within an individual, but also encourages accepting differences negotiated through family and society, inclusion of difference in what we value. What we value as diversity in botany and zoology may also be applied at the level of social engagements—face to face, mediated by institutions, and individual to group—and relations to power. Accepting differences will require effort to work around organismic self-protection. While *Selfie* encourages readers to understand how similarity scales throughout the biosphere, difference creates and scalar discontinuities abound. Matter was created by slight differences in temperature. Life was created by affinities of different molecules. Nearly every person was created by parents of two different sexes. Society is strengthened by the contributions of different perspectives, classes, races, cultures, genders, and skills. Difference creates.

These creative differences affect human biology so I tend to view my surroundings as outside, different, and potentially threatening. I more easily understand people who share my language and a certain threshold of cultural assumptions. For example, I have spent a lot of time working with poets in China where we focus on what we are doing together and how we feel in that moment. We frankly negotiate small differences regarding the poetry that we share. Both nations' poets tend to group by style and often those groups are named by the members or critics. The social dynamics of poets within groups are similar in both countries while poetry has a very different reception in the general cultures of China and the US. We also discuss how our different cultures handle gender which is a common thread, but race and politics remain tangential as a matter of politeness.

Closer to home, I sometimes find myself in conflict about when to support my own poetry and when the authors I publish. When I promote the recent Roof Books title by Will Alexander, *The Combustion Cycle*, I avoid including references to my own work in progress and accept the categorical separations of my different roles. As Alexander says:

> their breath begins to form as a transparent thievery / the fourteenth priority: consisting of grooming & violence the fifteenth: breathing simooms gauntlet & bribery.[2]

Ethnographically, social and political groups also tend to evaluate themselves as representing the whole. Many origin myths of early societies speak of the maker of the myth as "the people." The organization of 32 Tribal Colleges and Universities says that their:

> founders and leaders today believe that we are first and foremost human beings. American Indian tribes commonly call themselves The People, united by shared beliefs, customs, relationships, and a sense of belonging.[3]

Evaluating others as an evaluation of oneself is not only a psychological projection. It extends to group, psychosocial dynamics. Humans as a single species and as many interacting societies/groups have every incentive to improve our acceptance of each other and then of ourselves as we are, trending toward what we'd like to be. Because as climate change pressures our surroundings, social networks, and selves, we have no choices that I can think of beyond adaptation or significantly reduced population, geographical range, and possible extinction. Adaptation and significantly reduced population seem to me at this point a likely combination with some groups continuing to cling to assertions of primacy within the species.

But can groups adapt socially through culture and poetry to benefit humanity or does adaptation occur only through genetics and epigenetics of individual organisms? Social groups are heterogeneous, dynamic, overlapping, and even fractal and chaotic. Nevertheless, groups can adapt as organisms and organisms in many ways act like groups. Group adaptation occurs at several levels and is widely applicable:

- The individual cell is a group. Within the body, cells themselves are symbiotic groups of bacteria and native human components as Lynn Margulis showed in her paper on endosymbiosis, "On the origin of mitosing cells."[4] Chloroplasts and mitochondria adapted from independent bacteria to become part of eucaryotic cells to work together with other parts of plant and animal cells to generate energy for those cells. Margulis' idea has precursors, that is, it was adapted from earlier Russian ideas of symbiogenesis, a lineage and adapted, historical connection, if you will.
- Individual organisms are groups of these hybrid cells operating with non-self bacteria and viruses. NIH reports "the human body contains trillions of microorganisms—outnumbering human cells by 10 to 1."[5] If 90% of my body's cells are non-human (viruses, bacteria, and fungi), my body is a group, too. I can adapt epigenetically and behaviorally.
- The human brain according to David Sloan Wilson "is not a single general-purpose organ but a collection of many organs that adapt organisms to specific aspects of their environments."[6] Specialized mental circuitry helps birds learn to navigate, bats echolocate, and beavers extend their cognitive and physical security zone with dams and lodges. Culture and science writers have learned to distinguish the cultural understanding of mind from a more physical, material understanding of mind. Those looking at the problem culturally will tend differ from those looking at the brain's material aspects and interactions. The difference creates a conflict but accommodating multiple views of the same thing makes sense when understanding its identity, operations, and inter-intra-actions in multiple ecosystems of ideas and organisms.
- Many ideas of mind, one of the groups that changes social interaction, operate through different disciplines as an embodied entity (Lakoff & Mark Johnson), cultural phenomenon (Laura Dassow Walls), and specific psychological/psychosocial process (Sigmund, Freud et al.) The group of ideas about mind also operates through group interactions. Such classes of ideas often exhibit divergent vocabulary about the same component.
- Evolutionist Wilson proposes that all "three ingredients of natural selection—phenotypic variation, heritability, and fitness consequences—can exist at the level of groups."[7] So, natural selection alone occurs at the level of groups as well as the individual. Natural

selection also occurs in a group model with sexual selection as Darwin's late writing in *Descent of Man* proposed. Sometimes natural and sexual selection are difficult to separate. An inherent aesthetic tendency in humans and other animals such as birds derives along several channels from sexual selection. *Selfie* itselfie might be considered as a construction like a bowerbird builds.

- Social behavior evolves in a social manner in the way cells adapt through cellular inter-intra-actions. Each also change through exogenous connections.
- Allopatric (geographic) speciation is as important as sexual selection in evolution. Climate change will quicken and exacerbate the effect of this difference of location on the ability of species to reproduce.
- In *Descent of Man*, Darwin says:

> A tribe including many members who, from possessing in a high degree the spirit of patriotism, fidelity, obedience, courage, and sympathy, were always ready to aid one another, and to sacrifice themselves for the common good, would be victorious over most other tribes; and this would be natural selection.[8]

This is the kind of statement that led to Social Darwinism. In spite of the misuse of these interactions, cooperation at the core of the statement continues to support group success. Not only does cooperation support capitalism and capitalist expansion, mutual aid was also the ideal interaction for Peter Kropotkin.

> Besides *the law of Mutual Struggle* there is in Nature *the law of Mutual Aid*, which, for the success of the struggle for life, and especially for the progressive evolution of the species, is far more important than the law of mutual contest.[9]

Therefore, whether we're talking about Thomas Henry Huxley's *Struggle for Existence and Its Bearing Upon Man*, Darwin's scientific distance, or the political ethics of Kropotkin, no one, even Rand, ever postulated self-interest without cooperation. In Kropotkin, the combination of struggle aligned with what I have called the organismic binary, and mutual aid as a specific form of cooperation succeeds for people at all stages of development.

- Ron Silliman and Charles Bernstein, both successful (i.e., more widely read and influential than most other language writers) within the limited range of poetic success, were central to building a group identity for language writing. The origin of their common group identity is documented in *The Language Letters* book.[10] In the book, readers can understand how the ideas of language writing developed through the epistolary (cooperative group) form. Their collaboration constructs a group with connections between them constantly reaffirmed as they both respond to each other and initiate independent statements. Such connectivity implies an extension to environment(s).

  Silliman and Bernstein (with Andrews contributing the most important, if narrowly political, structural ideas about the program of the group, but he is not as widely read) act in the book as individuals focusing on the same set of targets, a group formation. The group also acted as an exclusionary organism, almost like a single author, evaluating who would be "good enough" in their terms for inclusion in the group and excluding the large variety of different tactics that were in use among language-centered writers at the time. After their group success, both writers attempted to build individual success through Silliman's blog and Bernstein's academic writing and organizing, but the group formation still dominates the public identity of both writers.
- Emily Dickinson, widely read after she lived, operated successfully, if peripherally, among Transcendentalists, particularly Ralph Waldo Emerson, who himself was influenced by Coleridge and other Romantic poets and linked to the transcendental philosophy of Kant. The Transcendental Club was originally formed in Cambridge by George Putnam, Emerson, and others. The transcendental voice endowed poetry with a sense of purpose. Dickinson benefitted from those connections and support from her niece Martha Dickinson Bianchi who inherited her manuscripts and revived interest in her writing. In our day, Susan Howe revisited Dickinson as an avant-garde poet, as opposed to or in addition to a women's writer with delicate sensibilities, in her works such as *My Emily Dickinson*.[11]
- Groups form, break down, and reform. Pound advanced Modernism through creating Imagism in association with T. E. Hulme and Hilda Doolittle (1912) and, a bit later, Amy Lowell. He then associated himself with other poets like Yeats (1913–1916) and Eliot (editing

The Wasteland in 1922). At the end of his life, after World War II, he met and influenced poets Olson and Allen Ginsberg who carried Pound's work into new groups.

- We improve our understanding of ourselves by group learning and practice. Wilson: "Our innate psychology creates a capacity for change by setting in motion a process of cultural evolution."[12] Cultural evolution lies at the heart of the processes that may be able to arrest global warming. Wilson:

  > Any unit [part of body, individual or group] becomes endowed with the properties inherent in the word organism to the degree that it is a unit of selection.[13]

  Cultural and biological change both occur through divergence.[14] (Elizabeth Grosz) Cultural evolution changes behavior, and that's what humans now need to do with our industrial practices.
- Cultural evolution looked at similarly to genetic evolution extends beyond biologists and feminist critics to anthropologists Franz Boas, Margaret Mead, and Marshall Sahlins in multilinear cultural evolution.[15] Donald Campbell coined the term Evolutionary Epistemology to establish the linkages between conceptual and biological evolution.[16] The risks of these linkages between biological and cultural evolution to citizens treated as biological experiments must be manifest to all. Nevertheless, mind is an outcome of individual biology, social engagement, and supportive surroundings, a group that drives change.

Groups, like individuals, adapt both selfish and altruistic behavior and combinations of each. Groups can support altruistic behavior such as alerting other members to dangers. They can also punish those who do not call out danger. Wilson:

> Punishers cause birds to issue warning calls that help everyone in the group, including free-riders who do not share the cost of enforcement.[17]

Punishment of individuals by the group through policing is a low-energy method to enforce altruism. Policing is not an arbitrary value, but itself a form that needs regulation like poetry. The social regulatory system is complex and layered with as much conflict about this subject among biologists as among social scientists. Among political leaders, conflicts in negotiations between oligarchic groups determine whether society provides more support to masses or to elites. Support changes dynamically. Pressure on oligarchic leaders helps modulate policing as Black Lives Matter shows through social pressure and the "defund the police" campaign shows through financial pressure. Groups adapt through these regulatory means as conditions change for individuals, societies, surroundings, and language as a specific set of connectors.

Groups add value to our self-image, reduce risk, and increase comprehension of our surroundings. Highly adhesive groups ranging from families to common phrases connect at multiple layers requiring less overt justification:

- Our birth parents have the same children or agree about prior children. They are united by law or common law into a legal family and often eat and live together as a family. They often share culture and share genes. As Chen Li writes, families scale: "A line of words forms a family in itself."[18]
- Words are located in phrases for efficiency and transport through subject-verb-object order and subject-object-verb order in sentences that organismic constraints predict. Most major languages follow some linear process like time or non-linear processes like manner and place. Michael Hahn et al.:

> ...language universals arise for functional reasons: that is, because they make human communication and language processing maximally efficient, and regularities across languages hold because these efficiency constraints are rooted in general principles of communication and cognition (e.g., refs....). Under this view, the various human languages represent *multiple solutions to the problem of efficient information transfer given human cognitive constraints.*[19] [italics mine]

The importance of communication cannot be ignored but other performative and presentational social interactions impact language and evolution as has been said earlier and must be emphasized throughout to avoid cramming all intra-action of the biosphere into a communications model.

- Phrase structure may be linked to structures in the brain that produce and use it. Brain activity. Matthew J. Nelson et al.:

> increased with each successive word in a sentence but decreased suddenly whenever words could be merged into a phrase. Regression analyses showed that each additional word or multiword phrase contributed a similar amount of additional brain activity.[20]

The last sentence indicates that the brain does not structure significant evaluative hierarchy for individual words, but does when understandable phrases are built. These methods for using words—grammatical rules and brain processes—help language communicate effectively by working with the brain. They also measure language.

Coherent, stable, and powerful groups have many links to body, social, and ecosystem processes. When the group adheres at only one or two layers such as a corporation, sports team, or poetry group, more explanations are invoked to justify group existence and strengthen the linkages:

- There are fewer connectors between individuals than within individuals.
- Common goals, theories, and rules are proposed through poetics, sports, and corporate identity to assert the specific identity of each entity.
- Explanations of origin, shared suffering, and common goals as in race, gender, and nationality abound and fluctuate according to the needs of the group.

These minimally attached groups have a greater likelihood of further fragmentation. Corporations, for example, have an average lifespan shorter than a person. In 1958 corporate lifespan averaged 61 years, but by 2017, a corporation's lifespan was reduced to a mere 18 years[21] (Dave West). These facts imply that weakly connected groups like corporations easily fall apart and need the support of other organizations like government or monopoly power in a supply chain to sustain them.

Literary groups are short lived as shown in Franco Moretti's *Graphs Maps Trees*. Picaresque novels begin to be written around 1750 and stop being popular when Jacobin, Gothic, and Anti-Jacobin novels begin to be written. They have a lifespan associated with generations, political, and climate events.[22] Language poetry lasted as a productive unit for about 20 years. As more information passes over group linkages, the more fragile they become, requiring more energy to sustain them.

Radical change through Marxist political formations lasted quite long, nearly 100 years. Yet the dialectic continues to retain value as an analytic tool. Factions form in those groups with fewer inherent connectors. Those factions tend to diverge:

- Identity groups often practice irreconcilable differences of language to help the groups cohere and sometimes to disguise their intent. In matters of goals and interpretation of information, identity groups within corporations vary based on how much power belongs to management, to labor, and to owners of firms. For groups of writers, use of language differs as a core dimension with political and personal implications
- Identities fragment as subgroups form around behavior, ethics, and distribution of resources for race, gender, writing style, class, and nationality groups. Subgroups, through personal relationships between individuals, develop an increased number of connectors within and between each other compared to the whole group. For example, class and gender often link to form a subgroup within racial identity that results in women being denied access to resources.
- Subgroups form attracting those who do not feel supported by existing subgroups. The Black Panther Party platform acted as a political alternative to NAACP efforts at integration. Subgroups form in a way that looks like biological evolution when the observer notices divergence within the group and dialectically when the observer notices how the new group emerged from the former group. Divergence first operates through one set of connectors and then emergence through another set of connectors: another syntax.

Weakly connected groups like churches, corporations, and nations hold together by encouraging communities to pass laws, service their members, and develop rituals to attract participants, protect assets, link to families (a stronger group), and establish morality to make up for their lack of inherent connection to the entire range of people's needs. Less sticky groups try to essentialize their connections with commandments, advertising/public relations, and constitutions, but usually points of view diverge as group interests spill into self-interest and are diluted.

When a group does not support the interests of all members, explanatory and justificatory connectors must be added to continue support of and by elites:

- Religion and culture sustain elites through talismanic texts, manifestos, positive criticism, attacks on outsiders, and artifacts such as poetry and music
- Within nations and corporations, other connections support elite interests. The group is invoked through constitutions, mission statements, and declarations of purposes and rights with always just enough promise or payment of cash, ideals, and promises to keep members' attention.

When one identity layer is continually threatened such as race or religion, group linkages operate unpredictably. Tensions emerge between allies such as between individuals with different roles in a racial community under a dominant racial leadership such white peoples in the US, Han in China, Hindus in India. A challenge to the Black community, such as the police killings of Black men, women, and children, brings out common anger and outrage at the police and attacks on their racist attitudes and procedures. Solutions to this outrage, however, appear through diverse sets of opinions based on the role of the Black individual in the community rather than a holistic persona or identity. The problem stems from reducing a person to their race: synecdoche.

On the economic level, the more workers seek to overcome their subjective and material alienation through building skills to increase their employability, the more alienated they become.[23] (Peter Bloom) Use of social media by youth increases their employability in technical jobs. Computers reinforce their use biologically through positive feedback and economically through alienation.

In a small non-profit, for example, the new manager wanted to sell t-shirts directly to members and subscribers even after being shown that t-shirt sales were minimal when made directly through the internet but increased ten-fold when coupled with a donation request. Increased donations occurred even when the donor did not accept the offer of the t-shirt.

Yet the alienated manager thought the t-shirt, not the donation, was the important driver of transactions with donors.

The presence of subordinate members in a hierarchy strengthens linkages within the group and binds the individual to the group. Leadership increases its wealth through the peaceful relations with other economic groups under capitalism. According to Thomas Piketty, workers' wages increase only as fast as economic growth, long term 1–2%, while capital returns 4–5% increasing the gap between rich and poor annually. When the gap becomes unsustainable, however, the networks of the rich are threatened by increased taxation of wealth, war, and revolution.[24]

Leadership also strengthens control over institutions by creating a conflict such as promoting hatred or distain of racial minorities by disaffected members of their own group to redirect the anger of that underclass from leadership and their abuse of power to a racial minority. The politics of resistance appears when exploitation by leadership provokes alienation from the group as much as economic alienation[25] (Albert Camus). The politics of resistance is a classroom for consumerism[26] (Pier Paolo Pasolini).

Similarly, but neither identically nor differently, as in the "Particles" chapter, pronouns conflict when someone is forced into a subordinate position or otherwise injured. The obvious grammatical example is the aforementioned use of the pronoun *he* as generic. Such a foray into questioning grammatical agreement creates conflict that highlights difference and at the same time establishes the identity of gender subgroups. In another example, if individuals identify with a noun, Latin languages can frequently cause offense in the way they gender nouns only sometimes for a comprehensible reason.

In a more complex example, on June 6, 2019, in the Democratic Presidential primary debates, Joe Biden said regarding working with segregationist Senators,

> We got things done. We didn't agree on much of anything. We got things done. We got it finished. But today, you look at the other side and you're the enemy. Not the opposition, the enemy. We don't talk to each other anymore.[27]

Other Democrats objected. Senator Kamala Harris pointed out:

> I have a great deal of respect for Vice President Biden. He's done very good work, and he has served our country in a very noble way. But to coddle the reputations of segregationists, of people who if they had their way, I would literally not be standing here as a member of the United States Senate, is, I think, it's just misinformed and it's wrong.[28]

How much conflict is generated by difference of opinion and how much by defending different use of words. Harris knows engagement is critical to the political process but wants to make her candidacy relevant by attacking Biden on the subject of race. Partially accurate interpretations of a speaker's intent, as in Harris' assumptions of Biden's intent, position her as a candidate. The proof of the need of a spectrum of intent is that Harris becomes Biden's Vice President. The language of conflict channeled collaboration that helps the Democrats gain the attention of voters interested in that conflict about collaboration.

Discussing the more general issue of conflict in climate change, Timothy Morton refers to universal ethics as hypocrisy in this excerpt from an interview.

> Everyone is a hypocrite at this point because of interdependence, because everything is interconnected. I can't get my politics and ethics completely correct. If I'm being nice to bunnies, I'm not being nice to bunny parasites. There's always something missing.[29]

Perhaps, as is the case with much of Morton's writing, this position is a bit overstated. Of course, we continue to try to preserve the lives of ourselves and our allies, but reductive thoughts like "everything has a right to exist"[30] cannot be the measure of social syntax. Social groups also have a right to determine membership. Reading the world forward, since that is the general direction of time, balance is functionally a useful target condition. Differences between forces created us in the first place, but balance and general consent provide in a wide range of target possibilities. Acceptance of common culture and law as opposed to focus on individual desires and self-interest accounts for many of the resolutions of hypocrisy. Rather than twist all decisions into value judgments that result in hypocrisy, I have been proposing taxonomy as an alternative to evaluation. The demonstrations of non-evaluative hierarchy (q-analysis) in subsequent chapters can be widely applied in informing solutions to slowing climate change and addressing concerns about hypocrisy.

Capitalist private surplus as opposed to common or community surplus requires continual expansion to be successful while environmental resource availability limits public surplus. The planet's carrying capacity of the immense load of humans now requires that we subordinate that capitalist humanist tendency to environmental balance and public surplus. But individuals and groups don't agree on what constitutes an appropriate target since a businessman can claim success by impoverishing families and despoiling his surroundings. We would be advised to avoid situations where effect comes near to harm and regulate society accordingly. But there are lots of contrary forces isolating factors from the whole.

Human suffering can be often stopped by addressing the cause rather than only helping suffering people. In the case of language, such caution should not really inhibit our genuine *freedom of speech* since freedom of speech in the US Constitution is intended for the purpose of speaking truth to power and not intended to harm fragile individuals or groups. Inappropriate use of freedom of speech to harm individuals with different opinions has been promoted by some US leaders as a way to create social change. But the only social change it has effectively promoted is increased control by those divisive leaders who themselves carefully monitor what they say to each other. Adherents to different perspectives often surprise me as in this example by an extremely popular Black detective novelist, Walter Mosley.

> How can I exercise these freedoms when my place of employment tells me that my job is on the line if I say a word that makes somebody, an unknown person, uncomfortable? There's all kinds of language that makes me uncomfortable. Half the utterances of my president, for instance.[31]

But how can equal and fair treatment of
others guide freedom of expression?
Both are now constrained by laws
affecting public firms.
Laws in an environmental era
must reflect those conditions.
Changing the ways that leaders view
themselves will help limit
the negative effects of inherited wealth.

Perspectives on Combining

"The 'personal' is already a plural condition. Perhaps one feels that it is located somewhere within, somewhere inside the body—in the stomach? the chest? the genitals? the throat? the head? One can look for it and already one is not oneself, one is several, incomplete, and subject to dispersal." *Lyn Hejinian*

Social Syntax

This chapter extends environmental ways of looking at the complicated notion of syntax. Rereading Aristotle's statements about humans as political animals, the chapter shows that commitment, reciprocity, and altruism combine into the cooperation needed to arrest climate change. While individuals, groups, and ecosystem can be defined, isolating individual ideas threatens social cohesion. At each level of the model, a concept like freedom works with other concepts: freedom and responsibility for individuals, freedom and equality for groups, freedom and the parameters of ecosystems for habitat, freedom and comprehensibility for poetry. Any ecosystem's network needs most of its many parts to function, prioritizing completeness over perfection.

* * *

Social syntax, how societies are ordered, can be read as an example of and isomorphic with evolutionary sequences like genomes and climate processes. Read in another direction, to survive the reflux after planetary ingestion of human toxins, readers benefit from allowing me to extend social theory into the language of the biosphere and investigate how it refracts in new habitats. The inside, outside, and borderlands of the model extend social structure to ecosystems. As Darwin says in *The Descent of Man*:

J. Sherry, *Selfie*, https://doi.org/10.1007/978-981-19-4870-1_8

> Now with those animals which live permanently in a body, the social instincts are ever present and persistent. Such animals are always ready to utter the danger-signal, to defend the community, and to give aid to their fellows in accordance with their habits.[1]

Individuals are more likely to survive and thrive in groups. Individuals engage with multiple groups through connections described in prior chapters. Individuals and groups also organize themselves and their connections in self-referential sequences. Group sequences range from the etiquette of conducting personal relationships and conversations, to our style of telling narratives, to methods of building organizations, legal systems, national defense, and protocols in negotiating societies, constitutions, and treaties. The order of social syntax launches and operates priority within groups, who goes first and why.

As Aristotle notes in *Politics*, not only.

> Man is by nature a political animal.

but also

> He who is without a city through nature rather than chance is either a mean sort or superior to man[2];

Reading beyond the initial period of identity (a semi-colon in some editions), the antihero and Übermensch reside together in one phrase at the edges of society, outlining much of the trajectory of social theory. But in the next phrases, the individual emerges from society, not the other way around:

> That the city is both by nature and prior to each individual, then, is clear. For if the individual when separated from it is not self-sufficient, he will be in a condition similar to that of the other parts in relation to the whole.[3]

The group is not a collection of prior individuals. Individual identities appear only when they begin to be distinguished from the prior assembled organisms at birth and then as children develop in families. Assumed syntax now roils with questions and instead of getting into the melee of these questions, let's frame the problem differently.

Emergence of individuals then becomes a question of scalar direction, now from many to one instead of one to many. Brian Goodwin: Evolution, by itself, described solely as species change emanating from inherited genetic change in individuals insufficiently explains the impact of chemical change, of the physics of rocks, of social change, of epigenetic change on evolution, and of climate change.[4] Certain individuals emerge to gain priority to reproduce, and their genome reproduces with indeterminate outcomes. The qualities that made them eligible to reproduce might not appear in their offspring.

More often, groups, rather than individuals, dominate to create imbalances in ecosystems like corporations in the six polluter industries today. Some groups dominate ecosystems and build stable relationships among multiple species like Brazil nut trees in the forest, aristocracies, and corporate leaders. Group actions using specific syntaxes, such as legal codes with hierarchies to keep order and connections to leading families may be able to determine outcomes, but often external events like drought play key roles.

For both individual and group adaptation, the climate of thought is ordered by conceptual ecosystems. Social media influence exemplifies current events. In small groups, individuals may have enough impact on the climate of thinking to make a significant difference. If a change agent who appears heroic tells a story about how poetry works in a way to make people feel good about themselves, they encourage interest and imitation. Changing focus helps understand how reactionary modernists like Pound and Stein inspired generations of Leftist poets. I might convince you that the innovations they made in form, method, and even style linked to an existing ethos of personal transformation in commercial cultures that attracted poets from the US to China for whom literary creation was more important than politics.

Cooperation facilitated by common assumptions in the climate of thought, will be a strong contributor to arresting climate change, stimulating social change, and moving poetry forward. But how is cooperation most readily realized? Altruism, self-sacrifice, and sharing are not adequately explained by individual psychology, game theory, and the visible shield of the organismic binary. Yet, models have been built to explain why human societies are so cooperative, and why human psychology seems to include prosocial motivations[5] (Robert Boyd & Peter J Richerson).

Altruism is a difficult topic especially in capitalist societies that lean toward individual freedom and the expectation of reciprocal support. At the same time, altruistic behavior may not be the most effective and certainly not the only way to promote cooperation. Incentives generated by and for the group matter especially for individuals who value reciprocity. Altruism and reciprocity are not so much additive but encourage different individuals to act[6] (Sung-Ha Hwang & Samuel Bowles). The environmental model asserts that commitment, reciprocity, and altruism together produce an inefficient but steadily progressive order of social evolution. Efforts to prove the value of one of these three mechanisms over the others to produce social goods have been the subject of many experiments. Rachel Croson prioritizes reciprocity from an economics perspective, but points out that each tactic has its own applications:

> Commitment has been used to explain behaviors like water conservation and voting. Altruism has been applied to explain contributions to charities and intergenerational transfers and bequests. Reciprocity has been invoked to explain gift exchange and labor market decisions.[7]

Since conditions and individual roles change continuously, the solutions applied by groups to execute change will consider commitment, reciprocity, and altruism in multiple interests, multiple personalities, multiple group needs, multiple habitats, and multiple situations of inter-intra-action. These many to many relationships mean that commitment, altruism, and reciprocity working together create the greatest chance for the cooperation needed to reduce greenhouse gas emissions.

A hard-nosed industrialist may suddenly become altruistic when a rare disease strikes their immediate family. An altruistic priest may not trust the stability of an arts organization in space rental negotiations and instead demand tough lease restrictions. My commitment to supporting poetry and poets over many decades is constrained by guidelines and incentives that I set with each writer I publish. A predilection for either reciprocity or altruism alone cannot gain wide cooperation even as one may be more important in a given situation.

More complexities! Cooperation is vital and risky, most widely used in one of several agreed syntaxes: mutually cooperative sequences where I do one thing and you do another to extend a process to a desired result. Cooperation can also be coerced in several ways through situations that stem from and yield bias for one group over another. Coercion channeled from bias promotes the values of the leaders of cities, reinforced and supported by those seeking to maintain privilege and stability such as rich people, old families, and the elites supporting them. These elites use policing as an inexpensive, widely employed, and problematic way to coerce cooperation in social engagements and illusory altruism in individuals. Here's how cooperation might interact with evolution. Nancy Folbre:

> Here lie some big questions for evolutionary thinkers: is racism within the U.S. a maladaptive preference—a by-product of the evolution of some other characteristic? Or is it a manifestation of intra-group conflicts with implications for a culturally constructed unit of selection known as race?... Cooperation itself is less important than the terms on which it is established and sustained. Hierarchical and authoritarian forms of cooperation succeed by distributing just enough benefits to the less empowered to leave them better off than their next best alternative. The only way to beat them is to improve that next best alternative.... That's what Black Lives Matter, with its emphasis on systemic change, is reaching for.[8]

Other examples of coercing cooperation include advertising, slavery, negotiating in bad faith to gain short-term cooperation, and withholding information about the extent of cooperation required by one party or the other. Groups like the State of Wisconsin under Governor Scott Walker coerced the cooperation of individual wage laborers with rulings where corporate leaders were allowed to collaborate but workers were denied collective bargaining[9] (Ballotopedia).

To undermine faking
and misdirection by leadership
cooperation works,
despite and notwithstanding.

In proposing and promoting process change to leaders of polluting industries, coercion has to be balanced with examples and incentives or corporate leaders will avoid change like any community resists change from outside to retain group cohesion[10] (Hwang and Bowles). In the 2019 paper "Business and the Fourth Wave of Environmentalism," the Environmental Defense Fund presents a 33-point opportunity gap where corporate leaders who support process adjustments for climate remediation are leaving money on the table instead of deploying cost-saving, environmental technology. While this paper from the Environmental Defense Fund presents only certain kinds of technology investments, readers can certainly see how prioritizing cost-saving sustainability technologies can apply at many levels.[11] *Harvard Business Review* points out that incentives to improve sustainability can also be applied to compensation.[12] Boardroom culture is already changing as I will discuss later in this chapter, albeit seemingly at a glacial pace.

Adding habitat to the social studies binary of individual and society, humanity's social conflicts can be less hotly debated due to the moderating effect of the third term: surroundings. Although more complex with the added term and considerations of habitat/surroundings, conflicts between individuals and groups will be easier to resolve because the analytic structure is more inclusive. Environmental planners need to seek the consent and collaboration of both leaders and specialists to assure that contributions proposed by any group can adapt to changing conditions. Altering the process of:

- Energy production by adding solar and wind to the grid needs excess supplies of fossil fuels during transition to renewable sources
- Transportation by substituting electric engines needs charging infrastructure
- Manufacturing by reusing inputs and recycling emissions needs technologies at all levels to work together
- Agriculture by recycling emissions from meat production and overflow of bio-stimulants requires large recycling plants
- Construction by adding low emissions concrete mixing needs the buyers of buildings to agree about prices
- Mining by recycling effluents and tailings can produce clay, tiles, glass, and concrete, but initially costs will be high.

Here's a simple validation of what societies gain from cooperative evolutionary group dynamics. Wilson: "Selfishness beats altruism in groups. Altruistic groups beat selfish groups."[13] Balancing differences between individual and group attitudes toward self-interest requires culture to change. In poetry, the emphasis on individual identity can change to include group input, group creation, and recognition of the chain of poetry sources, poetry creation, publicity, distribution, and the resulting climate of ideas influencing future poetry.

Variations in social syntax follow the Red Queen hypothesis: organisms adapt to (re)produce and to survive threatening conditions. As the Red Queen says to Alice in Lewis Carroll's *Through the Looking Glass*, "Now, HERE, you see, it takes all the running YOU can do, to keep in the same place. If you want to get somewhere else, you must run at least twice as fast as that!"[14]

Michael A. Brockhurst: Following the Red Queen's advice, some organisms change their method of reproduction from bi-parental sex to parthenogenesis when local conditions are safe, reverting to bi-parental reproduction (safer because genes come from more than one parent) when there are many threatening parasites in the pool.[15] Gender creation multiplies in safer cultures to show how the Red Queen vamps.

Carroll's Red Queen speaks to poets' risks as well as environmentalists'. Threats to poetry self-image, production, and consumption, like electronic media and internet discourses that achieve levels of surprise similar to poetry, drive poets to search for other workable syntactic structures, other reservoirs of meaning than surprise and excitement. Investigative poetry tactics like changing word order encounter difficulties similar to changing society and climate. Resistant, protective, compensating, and contrary behaviors all emerge from these levels of change. Syntax, that order of which I speak, applies throughout the environmental model in these isomorphic ways.

Holly Melgard's *Fetal Position*, for example, transforms found language and genre writing as a way of organizing labor—political, sexual, social, and hierarchical. The book really defies interpretation replacing it with taxonomy in chapters titled: "Reproductive Labor," where a woman talks about her ambivalence about giving birth, "Divisions of Labor," an alphabetized list of sounds uttered during dramatic portrayals of giving birth found on YouTube, "Child Labor," a pornographic monologue from the fetus' point of view, "Student Labor," a series of verbal gestures like excuses for not getting work done, "Lesser Person," a mother's loving and despairing description of her adopted feral child, and "Catcall," a satire of men harassing women written as a woman dog calling to an attractive yet somehow strangely diminutive man. "Hey pretty big little cute guy. Do you need a cuddle? Do you?"[16]

Without recontextualizing words, we're used to hearing in phrases, society stagnates through constant repetition. Through slang and genre, a person's identity becomes fixed, locked into these phrases. People will not easily adapt to the new information and stresses that arise in a changing climate and social conditions, dazed by devotion to individual predilection. Under conditions of stagnant syntax, parasitic behavior such as hidden bias in journalism, financialization, and autocracy proliferates because individuals are stereotyped as consumers, victims of weakly justified economic forces.

Can we adapt fast enough? Trilobites, even after millions of years of adaptation as Clark Coolidge writes, finally could not:

"ounce code orange
a
    the
        ohm
trilobite trilobites"[17]

Can our societies flooded with internet discourse raise their eyes from the digital treadmill long enough
to understand the problem
to act
to preserve the species? From Project Drawdown:

> …the notion of 'silver bullets' has persistent appeal—'what's the one big thing we can do?'—but they simply don't exist for complex problems such as the climate crisis. A whole system of solutions is required. Many climate solutions combine and cooperate, leveraging or enabling others for the greatest impact. For example, efficient buildings make distributed, renewable electricity generation more viable. The food system requires interventions on both supply and demand sides—e.g., better farming practices and reduced meat consumption. For greatest benefit, electric vehicles need 100% clean power on which to run. We need many, interconnected solutions for a multi-faceted, systemic challenge.[18]

This point becomes amazingly clear in the COVID-19-induced global supply chain disruptions of 2021. In this way, the environmental model changes people's collective mind and action, not only through what is presented in poetry, news, political speech, and scientific writing, but the orders and structures through which people use language. A purely political frame will fail humanity again as it, as in past eras, focused disproportionately on ideology. Cultural change mutates politics and connections between components of the biosphere suggest the method by which the logic of political economy produces phrases similar to biology.

In the same ways that bodies require a balance of nutrients, humans benefit by changing the social balance between individual and group dynamics more than seeking the political power many crave in the name of security. Environmental appeals to sustainability attempt to refocus culture away from competition toward cooperation and balance. A revolutionary replacement strategy will not be useful. Continual rebalancing rather than the wonderful thing, sleight of hand, or single lavish act serves environmental purposes.

For some poets, change means decentering reader expectations. Language writers, in particular, build ideas such as Mac Low's "reader-centered writing," that Hank Lazar explains seeks "a collaborative relationship between the reader and writer in the production of meaning."[19] Adapting language, treated as a tool, separate from the self, to an operational, collaborative component and building block of the world links with multidirectional connectors as well as its own and shared agency. Hélène Aji:

> If language is a medium, according to Bernstein, it is not the medium of our hermeneutic attempts, the medium through which we convey meanings about the world and our experience. Rather and more radically, it could be the medium through which 'we experience the world.'[20]

But I would not limit language to a medium. That language actually builds the world is not a new idea but a classical Latin phrase: *Lingua mundum aedificat*. Peter Inman moves toward that position by using gestural particles and partial constructions shuddering down the page in "Waver":

time as after
make in eye
(negged)
mulcher
dreyfusard
away at bends
(farm along by pea)
slowed blue off
cent quog[21]

Writers can help redirect how people develop themselves in a similar way. Leaders of industry, global resources, and the means of production can no longer be viewed as simply instrumental. They, like language, are creating the world climate and determining much of our thought about it. We, the people, and they, the leaders, must be viewed as components of each other, not merely antagonists.

For minority communities, such change means asserting processes that provide them with the ability to refer to themselves and act for themselves through methods similar to the dominant race in any jurisdiction. This can be true for black people in the US, non-Han people in China, and different ethnic groups within African nations. Delicate attention to immediate details removes the poet from the conflicts of power into an integrated relationship to their surroundings, but does not provide access to the levers of power. A long supply chain is required for poetry to influence thought, but those connections exist.

Perversely, institutions wielding power and poets wielding tradition promote the organismic binary as an essential value and identity often to the exclusion of collaborative efforts. Further, change as balancing has often been associated with allowing the rich and powerful free rein through laissez faire economics and oligarchic control. Re-balancing society for individual and group improvement, in the sense of including habitat in the social formulae, conditions writers to adapt beyond aesthetic practices. Phrased this way, a difficult task inspires poets to adopt channels of writing that have been used primarily for social regulation.

Another way of looking at the history of literature is to view it as propelled by writing as dissent. Even as it was focused on politics, language writing in Silliman's words established an environmental model of change.

> Poetry was destined to change not merely because the content of daily life had changed, but also because the makeup of possible audiences was no longer the same.[22]

Including the ecosystem *of* poetry *in* poetry was an important impetus for language writing as was the epistolary process that started it as a coherent concept, broadening the reach of those ideas through publication and building audiences and institutions. The notion of language writing as social critique was a brave folly based on the very essentialism of language it criticized elsewhere. Yet, progressives of all stripes flocked to poetry's leftist pretentions. But it must be repeated that language writing was also an ecosystem defined by exclusions as has been the case for most writing "schools." L=A=N=G =U=A=G=E magazine, for example, did not provide fertile soil for writers focused on New Critical writing where the poem itself was the center of the poets' investigations.

In another example of exclusion, Armantrout, one of the writers near the center of the language group, wrote an essay titled "Why Don't Women Do Language-Oriented Writing." Armantrout points out that, "Most male writers aren't language-centered either." Then she says of language-oriented writing,

> I use that term but I'm suspicious of it, finally, because it seems to imply division between language and experience, thought and feeling, inner and outer. The work I like best sees itself and sees the world. It is ambi-centric, if you will.[23]

The target I call environmental writing—poetic, critical, and hybrid—continues to shift so thinking about ecological processes through description of place must be carefully rendered to avoid inhibiting change by modeling permanence. Describing change in static units discourages readers from accepting change. Rather, description reinforces the organismic binary of self and other as well as existing standards even when seeming to be workmanlike, stolid, and sincere. Gary Snyder:

Lay down these words
Before your mind like rocks.
        placed solid, by hands[24]

Both ekphrastic and conceptual writing draw on pre-existing texts in order to "innovate." Through a parody of convention, they establish a platform from which to change, but their process, drawing inherently from one source like a painting, inhibits change. Conceptual writing even can be viewed as mocking aspirations of communities that demand change as some of the writers of that group discovered to their unrepentant horror. One can say, however, almost all writing draws in significant ways on pre-existing texts and methods, so the issue reflects from many facets if turned into an argument.

Applying a received syntax of social interaction (business as usual, top-down oligarchy), seriously, since politics has little sense of humor, to climate change, however, disregards the increasing rate of change in the biosphere that whisks humanity toward change imposed by the planet in reaction to human insults. Nevertheless, most environmental action will be accomplished by the same small group structure that I critique as oligarchic. Writing needs new orders of social interaction to support its readers. Historical and received views of nature, such as *man's* domination over the beasts in Genesis, Buddhist hierarchies, and Plato's view of nature as clarifying confused human perception, are likely to reproduce the conditions and behavior that drive climate change in the first place.

Applying received language to every situation without playfulness and awareness of the legitimate stresses and conflicts that created them risks misrepresenting, offending, and simply being wrong.

> He's a slut. She's an asshole. Republicans only support the rich. Democrats only tax and spend. Language writing is incomprehensible. Conceptual poets are racists. NY School poets are narcissists. Simple understandable poetry is democratic. The biosphere is too big to affect.

In today's polarized political climate, received history and the organismic binary tend to separate freedom from other concepts and sequences of the social syntax. Isolating the idea of freedom breaks down social cohesion and shakes the foundations of consumer society that themselves support isomorphic freedom.

Freedom only functions effectively when constraints like individual responsibility create an operational and fertile pair with freedom. In group formations, freedom pairs with equality to maintain stability. For ecosystems, freedom must be balanced with the needs of habitats or unsustainable pollution results. For poetry, as mentioned above, comprehensibility constrains free use of language and form.

In the political landscape, generalizing about freedom and isolating the concept from its ecosystems has created a danger to general consent. As Latour says, "we must resist such simplifications with all our might."[25] Freedom and the expectation of freedom differ depending on which nations, cultures, and communities an individual inhabits or adheres to. Chinese freedom of action differs in scope from American freedom of action, creating different writing. In my work with Chinese poets over many years, I find that the Chinese poets expect, although they're not all happy about, regulation by government of publishing, speech, and access to information. Chinese contemporary poetry adapts to this political situation by using an old Chinese tradition of metaphor to create scalar connections between, for example, a description of animals and a critique of leadership's policy of wealth extraction or really any exercise of power. Che Qianzi:

I rented housing at the buffalo's back, a bit far from home,
water buffalo, lean as a few bones wrapped in newspaper.

And fat sheep,
wool plucked clean,
only the size of frogs...[26]

The US poet expects to access information and speak freely, although in reality government and information media obscure important information so free speech in the US is conditioned on available resources and vary widely depending on the sources used by the speaker. US poetry is more frontal when speaking about power relations. At the same time US poets, due to the writings of minorities of race, gender, class, and ability have become more careful about speaking about other people lest the poet insult others by assuming knowledge about another person. The characteristics of connectors and access to media between specialists and elites determine prerogatives in both nations. Differences in poetic address and differences in government regulation dominate public discussion about the poetry of both nations, but similarity in the connection between classes of readers and the taxonomy of ideas make conversation among Chinese and US poets fruitful.

Socio-centric models like popular and academic views of Capitalism can no longer govern effectively alone because it curbs interaction rather than actually supporting open markets. Ferdinand Braudel suggests that Capitalism fosters restricted, not free, markets where prices are controlled by "power and cunning." In Capitalist markets, "everyone would be sure in advance, with the benefit of common experience, how the processes of exchange would operate."[27] Capital fails to produce general consent under volatile conditions as we see in discussions about individual and political freedoms. These failures would multiply quickly when climate deteriorates. General consent around environment is a serious threat to elite power. Latour:

> The obscurantist elites, according to this narrative, have taken the threat seriously; they have concluded that their dominance was threatened and have decided to dismantle the ideology of a planet shared by all; they have understood that such an abandonment could under no circumstances be made public.[28]

This statement fortunately is not universal. Recent statements from major financial institutions in favor of Stakeholder Capitalism such as Black Rock's Larry Fink support consideration of customers, suppliers, workers, and investors—the entire corporate ecosystem in corporate strategy.

> 'In today's globally interconnected world, a company must create value for and be valued by its full range of stakeholders in order to deliver long-term value for its shareholders.' Truly great companies, he continues, share 'a clear sense of purpose; consistent values; and, crucially, they recognize the importance of engaging with and delivering for their key stakeholders....' That view stands in sharp contrast with the idea that the role of the corporation in society is simply to make money for its shareholders by whatever legal means necessary.[29]

Sociocentric models also become tangled when we need to understand connections from society to habitat to reduce pollution and maintain access to resources such as food and energy. Capital often obscures those connections because they would lower profitability. Latour:

> Before accusing 'the people' of no longer believing in anything, one ought to measure the effect that overwhelming betrayal on people's level of trust.[30]

I don't expect corporations to suddenly stop seeking wealth, but changes such as Stakeholder Capitalism do presage a more environmentally sound structure for production that feeds, clothes, shelters, and engages the world.

* * *

Appropriated language and unrestricted freedom are not the only questionable rubrics some writers apply to poetry. Those who fight for clarity of writing at every turn are frequently surprised, frustrated, and ultimately depressed by non-linear reality, its swerves, and the need to invent to keep up with, not just unpredictable individual psychology and sensations in the body, but also fluctuations of politics, culture, and habitat. Rather than transparency of language, as an editor, I insist on a constraint that even when writing is semantically difficult, the writer's intention must be comprehensible or no one will read the poetry.

The links to other ideas, other poetry, and the frames of poetic intent benefit from pointing in directions the reader can follow, even if the traffic is congested. Environmentalism frees poetry from some prior limits of form and address, but also establishes new boundary conditions. The goal posts are moved and the boundaries will benefit from full exploration as opposed to imposing narrower restrictions. Poetry may still, I hope, afflict the comfortable.

Environmentalism shows how meaning travels along specific although complex routes. These limits channel particular options for meaning, especially connection and intersection. The term Intersectionality, an obscure legal concept coined 30 years ago by Kimberlé Crenshaw, has become a general statement about the network of bias for race and gender. In *Selfie*, I model the network to describe the both general and specific channels of inter-intra-action. Exploring those pathways appears vital for poets who want to participate in a larger frame than poetry as a performance of self, sociocentric writing, and poetry about poetry alone where routes are well mapped and frankly risk more than writing into rough terrain.

Since most writers know that it doesn't make sense to write down any old thing that comes to mind, practitioners apply several kinds of attention to their process, shapes, and themes among which are methods that societies and ecosystems also use. With new ideas and solutions cropping up through both novel and well-understood channels, a range of possible solutions act as a reservoir of writerly resources that is more sustainable than both unrestricted freedom and rigid order.

Although both freedom and order have produced brilliant results, the coexistence in most decades of canonical, new, and hybrid poetry matches actual and perceived events more closely than the polemical language that triggers narrow emotions and focuses on the conflicts and their binaries that force readers to evaluate a poem like they evaluate themselves. Many readers like this reinforcement of their organismic binary, but the world is not only that binary. In a continual negotiation across ecological networks, some styles dominate time periods, but species thrive and recede with fashion, a supportive ecosystem.

As an activist, refracting the world solely through the organismic binaries of oligarchic self-interest and writerly self-absorption reduces my ability to have significant impact on conflicts and threats of climate change. Therefore, social syntax needs multiple inputs from habitat. Scientists have suffered from the same blindered exploration. Hugh Patterson: Had Darwin in *The Descent of Man* not "abandoned environmental selection in favor of competitive selection, and adopted relative adaptation with all its consequences as an alternative,"[31] we might have a scientific community that recognized not only factual truth but also the practical realities of human misperception about global warming. While Darwin provided two different perspectives in his two main books, Alfred Russel Wallace also maintained that several environmental factors determined fitness as well as fitness of individual organisms and their social interaction[32] (Michael Bulmer).

Group behavior slows and speeds climate change. Social dynamics, such as promoting smaller families and corporate activities that include all stakeholders and cleaning up after industrial extraction, gain greater traction by including group evolution and group connections metaphorically and in actual plans and processes. Building an overall ecology of interactions evolves individual and group culture toward inclusive practices.

Environmental pecking order—
personal, social, ecological,
connected with syntactic similarity
(summer, fall, winter, spring)—
plays with poetry through conditions
of metaphor and form,
changing like organisms
evolve by divergence,
evolved with streams of themes
refracted through taxonomy.

Society builds repeatable
processes like law and
manners through institutions,
changing, too, by divergence,
by sharing common processes.

Diverging from received syntax and opening new ways of
looking at exceptional and commonplace connections to
politics can save us from banality.

The sequences are seemingly
  unbreakable chains of seconds,
which the surroundings demonstrate
best when left alone,
and so I breathe
the smell in the
dark prop plane, feel
the motors and hear.[33]

Perelman's change of syntax turns on "and so I breathe," applying non-standard syntax in writing, society, and surroundings that both risks misunderstanding and presents opportunities for change. I suspect he kept his seatbelt on during the flight.

In humanist terms, we apply technique to make the new comprehensible. Put environmentally, non-standard syntax supports a different range of comprehension than the usual ways we use language. New ways to understand things materialize out of expanded syntax, but also increase the number of ways of making meaning that readers can comprehend. I am sad to say this, but simplification, with its complicated isolation of freedom from freedom's ecosystem, has, in the most dangerous manner, obscured our ability to see and comprehend our world.

How Can Culture Change Habitat?

In this chapter, I discuss how syntax change might work when readers accept that the development of language operates through the order of both societies and ecosystems rather than an only private sequence. Old habits die hard even among environmental organizations. The chapter shows how the mechanisms of change operate in individual behavior, group activity, and ecosystem responses through current events, social media, and poetry. The chapter shows how fundamentally different ideologies exhibit similar dynamics of change. The chapter presents an example of how proposing radical change makes intermediate changes more palatable to slowly changing culture.

* * *

Unless we are prepared to allow civilization to collapse before building an environmentally modeled society, we benefit by convincing large groups of people of the value of paying attention to how climate is changing and changing how they order their lives and minds. Changing production goals to prioritize sustainable methods without also changing the cultural and political pressures on leadership will tend to reproduce current conditions. Problem/solution models that initially drove extract-and-abandon linear practices—evaluative hierarchy, isolating efficiency, and separating resource practices from consequences, self-aggrandizement, and private accumulation—can be replaced with cyclical, self-renewing processes.

J. Sherry, *Selfie*, https://doi.org/10.1007/978-981-19-4870-1_9

Mainstream media presentations using cyclical models quickly change attitudes when presented with specific intention. The new ways of engaging poetry and other uses of language in ecology include but are not limited to new forms, new theories of language, and greater political involvement by poets.

I'm not suggesting that old methods of engaging events, institutional boundaries, and cultural connections are easy to change. In 2013, for example, a group of friends from Occupy Wall Street approached 350.org with an example of how institutional endowments might eliminate fossil fuel producers from their portfolios without losing income. The new indices returned nearly 0.5% more in a 20-year regression analysis than the old ones that included fossil fuel producers. But leadership at 350.org did not want to adopt the new fund listings because they did not want to play favorites by selecting only one provider of indices. They were offered the fund methodology for free but they refused to become involved in this part of the process, saying it was the consultants to endowments whose job included figuring out what instruments to invest in. 350.org never seemed to understand how evolution worked by introducing a single change to see if it would make a difference, establish an advantage and then reproducing it. This excellent and vital organization could not or would not understand the value of vertical control in implementing change. Instead 350.org insisted on capitalist models of separation and division of responsibility to which they were habituated. Their ability to increase divestment was hampered by their attachment to cultural standards.[1] Adapting is difficult and essential because as Jem Bendell points out: "Some of us may take pride in upholding the norms of the current society, even amidst collapse."[2] Stiff upper lip, chaps; Bendell has few supporters among scientists.

More effective management of how end-to-end business processes are modified might include changing together what operates together in an industrial, social, or even poetic process. The melding of poetry and criticism in language writing by linking non-academic poets to university English departments marks such a modification in the literary world. Forming an appropriate taxonomy of these manageable groups of methods supports knowing where and how to exert control as well as how control operates in the biosphere. For example, change:

- from criticizing how we treat our surroundings
  to including our surroundings in definitions of ourselves and societies[3] (Moore)
- from holistic goals for individual and group identity
  to performative, componentized views of who we are
- from the received syntax of political and poetical leadership
  to increasing awareness of the separation of form of governance and poetry like democracy and authorial focus on poems from their operations like oligarchy and end-to-end poetic process
- from increasing refinement and complication of past modes of expression in order that new events can be accommodated
  to modifying the syntax and semantics of writing through experimentation and innovation to look and function more like the goals of behavioral, social, and evolutionary change.

The value of changing received syntax, including the order of phrases, appears as soon as population and other conditions begin to change, which turns out to be all the time at different speeds. This raises the question of what velocity of change (acceleration) is the threshold required to create awareness of change within different societies. Within certain ranges change doesn't matter that much. Readers and citizens simply accommodate it.

The change from Metaphysical poetry to Romantic poetry in English occurred during the long period from the death of Pope in 1744 to William Blake's *Songs of Innocence* in 1789. Interim writings contain components of both styles such as Thomas Grey's "Elegy Written in a Country Churchyard" (1750). In Germany, the immense and immediate popularity of Johann Wolfgang von Goethe's *The Sorrow of Young Werther* (1774) made for more radical change and influenced English poets. George Gordon Byron loved it and Napoleon apparently read it seven times.

In my lifetime, some quick and some slower changes occurred. The sudden appearance of Beat Poetry at the Six Gallery in San Francisco (1956), the publication of Ginsberg's Howl (1956), the ensuing obscenity trial (1957), and publicity around Kerouac's *Dharma Bums* (1958) all coming in a few short years increased awareness about poetry in the world and initiated, supported by the Baby Boom population bulge, what came to be known as the "counterculture." Language writing and its offshoots changed the poetry world over a generation. Older traditions didn't disappear but the new framework dominated discussions among young writers and then in universities.

The first wave of "language-oriented" writing started in the Bay Area in 1971 with the publication of *This* magazine, edited by Grenier and Barrett Watten, and other publications edited by Hejinian (*Tuumba*), Perelman (*Hills*), and Silliman (*Tottel's*) as well as readings and talks. The second wave started in 1976 with my own *Roof Magazine* linking NY School and Language writing. Early the next year, L = A = N = G = U = A = G = E magazine edited by Bernstein and Andrews pushed poetry into a hybrid poem/essay model, a major change that significantly expanded the range of tendency from the counterculture to university settings. Dozens of publications and the takeover of many university English departments by Language writers stemmed initially from the SUNY Buffalo Poetics Program. Initially, in collaboration with Albert Cook, who thought poetry and poetics could be taught by non-academic poets, Robert Creeley sponsored Bernstein's hiring as a tenured professor of English with only a bachelor's degree based on the strength L = A = N = G = U = A = G = E magazine. The poet theorists who emerged from the program, along with the first generation of language writers, raised awareness of the poetry and ideas to an international scale. Since then, several other tendencies have mutated language writing and the poetry world's perspectives.

Change outside expected ranges, such as introducing a new idea in poetry, a new technology, or an external threat like a tornado and then suddenly operating around a new attractor like language writing challenges the ability of individuals and groups to adapt and survive. Poetry, for example, suffered initially from introduction of electronic media with both images and text on the internet, but adapted with a huge increase in MFA poetry programs in universities from 64 in 1994 to 229 in 2020 plus an additional 152 Masters programs in creative writing,[4] increased independent publishing activity, and public readings both in person and more recently online. New technology drove poetry in new styles such as Flarf which selected its vocabulary from internet searches, but presented through older methods like formal education, public readings, and book publishing. Some new formats of writing were developed online, but few compared to the resurgence of older styles such as field poetry and other twentieth-century forms. These events, retreating to older form when threats appear, are good examples of the reactions of risk homeostasis.

Through the internet, change can be made to happen somewhat differently. In their talk, "Hacking Ten Million Useful Idiots: Online Propaganda as a Socio-Technical Security Project,"[5] Pablo Breuer and David Perlman examine how rapid change from norms can be achieved with social media. Given a set of norms of acceptable discourse, known as the Overton Window, around climate change, industrial process, capitalism, and their interactions, Breuer and Perlman propose a point of view that's way out to the far right on the political spectrum that a politician or foreign government wants to project to Americans.

It would take a lot of work and a long time to use traditional public relations to make this fairly radical notion acceptable. Instead of the slow method, they would start with a "crazy pitch" and rally a small, vocal fringe to the really outlandish notion that climate change is a fiction invented by the Chinese Communist Party to destroy capitalism. If Trump can get enough people talking about it, his foot is in the door, and the Overton window stretches to include both the wacko hypothesis and his fairly radical target policy such as reducing funding for clean air and water. The target policy suddenly seems sensible compared with the wacko opinion and people now consider it a viable alternative. The "useful idiots" have spread Trump's radical policy by talking about the wacko opinion that China invented global warming.

Traditional methods of fact checking such as *Politifact* take a long time to verify—too long for the trajectory of internet opinion. Plus, automated systems don't handle satire or editorial content very well. Using propagation-based techniques bots continually rephrase the false proposition keeping it visible to large populations, so while truth is accepted, included, and forgotten, monstrous lies are reiterated constantly.

In the scalable networks of climate systems and other non-linear processes, people "form network links at a much faster rate than you lose links."[6] The governing principle is not creating the link to the false information, since you can see how easy that is. The problem is losing the link, untying dyslexia, erasing an impression, and becoming convinced that what you heard, connected to, and believed or even considered is not true. Such tools can be used for multiple purposes. The ethical question is whether to use such tools to promote environmental targets. How much exaggeration makes sense; how much engages falsehood in a way that turns environmentalists into autocrats and polluters?

Let the reader judge this extended and perhaps wacko connection: Labor and material resources are the primary values input to an economic system. But treating labor as a resource like energy and iron misrepresents workers as passive things: their welfare can be decentered by ignoring external costs such as health care, breakdown of family structures, and wages driven lower, since lowering prices of resources is a primary method for industrial and financial interests.

The labor/resource identity falls apart easily under environmental scrutiny because it poses well-known risks to workers. As an alternative, many in the environmental movement decenter humanity. Does that imply that the environmental movement might be willing to use labor as a resource to manage without regard to worker's well-being? One solution is for the decentering, as it operates within neoliberalism, to be replaced by pairing human labor and habitat welfare together in industrial processes such as proposed in the Green New Deal. If leadership models composite things through connectivity, then what they've learned from industrialization can be applied to correcting its destructive apparatus.

If I apply this environmental economics of supply and demand to poetry, does it increase the value of poets who cross disciplinary boundaries or will I reduce the inherent value of each individual poem? Certainly, the poem fits into a larger structure of society, language, and management of language. As an analogy, increasing the number of books in a print run decreases the cost per copy. But does the presence of more poetry available for readers decrease the value of the individual poet and poem through the law supply and demand? Or might the extra energy applied to poetry travel through connections that increase its overall audience and hence the value of poetry for all of us? Asking the question this way shows how the numbers of our "sweet science" measure multiple, additive paths, not only the zero-sum fallacy of commerce.

The supply and demand model also increases stress on the ecosystems contributing poems unless you also calculate the value to readers of poetry. More people writing poetry and reading poetry broadcasts consideration of poetry throughout the network of readers, writers, critics, editors, publishers, and distributors. Increased traffic in any part of the poetry world—more poems, more readers, better distribution, visible influence of poetry on other parts of the culture—changes ideas about poetry and this can create new linkages to other cultural-environmental ecosystems. For example, when language writing refreshed ideas around poetry, the increased number of students increased the number of books sold and read. Language writing, importantly for this book, increased the number of connections to other disciplines such as science (Armantrout), philosophy (Bernstein), and politics (Silliman).

The attention in recent years to marginalized class, gender, racial, and ability groups has created new poetry readers excited to hear about their peers with whom they identify. This focus also encourages many poets to write again about their individual experience diffracting the kind of social and environmental progress made by language writing and ecopoetics. Expanding the numbers and types of poems available—language writing, ecopoetics, identarian—increases readership and separates groups. Groups' wider distribution increases their chances of survival, so expanding readership around the globe sustains styles and methods of writing.

The argument for increasing the number and frequency of building blocks of poetry—poems, books, readings, publishers, and so on—encounters limits within literary ecosystems. Without new attractors like new forms and methods of writing or giving a voice to underrepresented communities that bring new readers to poetry, there is a limit to the amount of poetry that will be read widely: a glut of poetry books may then decrease the value of each book. One could imagine a bell curve where the value of poetry to society increases with the number of poets, poems, and poetry books until the saturation of potential readers begins to decrease the value of each poem, poet, and publisher. New ideas about poetry build new linkages to other nodes in an ecosystem.

Looking through the lens of environmental connectivity that poetry shares with other cultural categories, similar processes operate at different scales including how much and what kinds of food are optimal to eat, how much of the economy can be managed by government regulation, and the carrying capacity of an ecosystem for any one species. Here's the odd connection that also acts as a metaphor: The Armey Curve in Fig. 1, shows a related growth pattern under different levels of government spending.[7]

The Cato Institute would have government spending under 25–30%.[8] The Armey Curve shows that developed economies average a higher percentage of government spending than emerging economies. Too much government spending removes incentives to innovation while too little fails to build collective infrastructure with predictably slower growth and unequal income distribution.

Similarly, green technology to slow climate change will increase government spending as a percentage of nations' GDP, but limits will

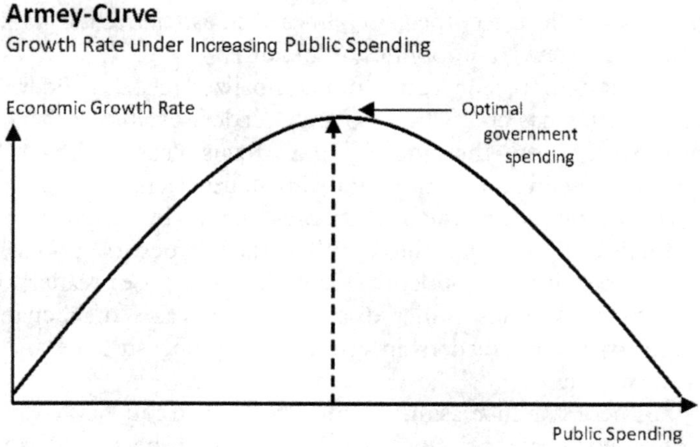

Fig. 1 Armey Curve of Growth Rate under Increasing Public Spending

make themselves known in slowing both growth and inequality as in European social welfare economies. Fortunately, there is little risk of a glut of poetry, since unread writing always exceeds well-thumbed volumes. It's the usual state of poetry. Have I reached beyond acceptable limits of argument and is the value of this discussion now decreasing?

Actual events also change people quickly. Can we acknowledge climate changes are already occurring or do our lives have to be upended while we wander about wondering what happened to Katy Perry? Multiple methods may be engaged. When asked to give a reading, I hungrily fly in a jet plane to California. I'm not promoting indifference, but the pabulum of greenwashing beckons from the shoals with its easily digestible reassurances. For those paying little attention, greenwashing may condition people to accept change. On the other hand, solutions that link poor people to green jobs connect social benefits and labor progressively so workers in both government and manufacturing can learn to operate together on unfamiliar terrain.

Poetry can inscribe and voice new associations. Positioning people to change their syntactical links, even when their long-term survival is threatened, requires leveling the field of expectation through widely heard individuals, leaders, and speakers for each group. Perhaps that includes engaging some wacko points of view to make moderate change palatable. Many methods appear feasible helping to create value for different groups.

Now that a selfie with a comprehensible background makes crossing borderlands seem productive, look at poetry another way. Does the focus of poetry on self translate to high public value and public agency or does it reinforce consumerism? Do poets lose the ability to influence their societies and their surroundings as they deal with themselves? To judge from the success of Amanda Gorman's expressive poetics in reaching the presidential inaugural podium in 2021, including oneself in polemical poetry does well in America! Oneself and one's feelings may follow the Armey Curve, expressive writing rising and falling with the breath of social discourse. Breath and sound are just part of the process.

Romantic poetry, from its beginnings in Goethe to recent Modernist versions, reinforced consumer capitalism and perhaps enhanced it. Romanticism both heightened awareness about industrial destruction in poets like Blake and was co-produced by industry in modern writers like Stevens.

Complacencies of the peignoir, and late
Coffee and oranges in a sunny chair,
And the green freedom of a cockatoo
Upon a rug mingle to dissipate
The holy hush of ancient sacrifice.[9]

The multidirectional interaction continues to be related to how surroundings affect the poet and also how poets as a proxy (and, in fact, as a force in themselves) affect their surroundings. In postmodern humanism and environmentalism, the answer has been that improving the understanding of what it means to be human through poetry creates public value. "The personal is political" remains a fine example. This phrase fortunately has collective authorship and mutual influence according to feminist writers like Shulamith Firestone, Robin Morgan, Kerry Burch, and Carol Hanisch who wrote an article with that title in 1970 but disavowed authorship.[10]

Some poets intentionally and directly claim to change civic discourse and the terms of public engagement in order to improve the condition of individuals oppressed by political systems. From Catullus' attacks on the behavior of political figures to William Shakespeare's cheerleading for the English monarchy to Amiri Baraka's critiques of US military expansion, poetry engages multiple discourses through specialized syntax and diction that includes figures of speech oriented to public themes.

If poets write to influence politics, do they lose their connections to themselves? What if influencing politics builds a new set of connections without eliminating prior channels even though they may be less traveled? Such a high level of maintenance creates anxiety that threatens the social media consumer if they lose contact for even a minute. At a little lower layer, the linkages that connect the individual to other entities within ecosystems operate as analogs to each other. They also develop material relations. Isomorphism, where it operates, might make any theme that's carefully constituted analogous at other scales. This inter-intra-action, performing both between and within subjects, confirms the connected identities of human and non-human components of the biosphere as well in poets as diverse as Olson and Francis Ponge and among contemporaries such as Mei-mei Berssenbrugge and Armantrout:

The child wants his mother
to put her head
where his is, see
what he sees.[11]

Armantrout affirms both the value of and desire for confirmation. Berssenbrugge builds connections between the poet and plants, writing that "The rose communicates instantly with the woman by sight, collapsing its boundaries, and the woman widens her boundaries."[12] Many people desire a common point of view or subject position as Armantrout suggests. Going further, Berssenbrugge merges subject boundaries to show consciousness at different levels can still communicate. Then, Fred Moten pluralizes the subject as a language trope from a subject (noun), to subject (verb), to eliminating the subject in a free flow, free of subjection:

> Indeed, our resistant, relentlessly impossible object is subjectless predication, subjectless escape, escape from subjection, in and through the paralegal flaw that animates and exhausts the language of ontology.[13]

All these poets work toward a new, combinative subject in support of the notion built here: changing how we think about ourselves remains critical to slowing climate change. Discussing the other side of the coin, identity of definitive subject action, Reilly quoting Cai Guo Qiang's essay reveals the negative impact of such composites:

There are more pigs
than fish in the Jiapingtang River[14]

Humans continue to be increasingly subject to the vicissitudes of difference when we ignore our common location and responses. Not everything scales and the subject also retains its liminal and physical identity, but scalar metaphor is a unique and powerful tool against climate change. Community analogies described by poets concerning class, location, race, ability, and gender also establish these composite entities with material linkages like language references and constructs, cultural artifacts, and performances.

These connectors are not my invention. Rather they have been suppressed in favor of poems and poetry as things and in themselves. New Criticism, a method of reading that ignores the text's context, author, and surrounding environment, promoted the folly that all meaning of the poem resides in the poem itself. Language writing, centering meaning in and fetishizing language and language interactions, has largely begun to identify its own intra-actions as such. For language writing, in its internal relations, its equal signs, if you will, language itself acts both as an interconnected thing within itself and a connector to the rest of its outside themes where value of the latter shrinks to inflate the former. This connector model of language writing can be generalized to all avant garde, experimental, innovative, and investigative poetries. This generalization, both factual and conceptual, differs from simply readmitting the referential into the subject-decentered poem because reference both exists at different distances and moves toward disappearance. There is only an asymptotic approach to "non-referential writing" as much as Grenier used that phrase to good effect.

Experimental, avant garde, exploratory, and innovative poetries have built discursive connectors:

- people, politics (conflicts and alliances), and language (language writing)
- people and prior writing (ekphrasis, creative criticism, and conceptualism)
- individuals and other components by using procedures (Dada, Oulipo, Fluxus, Flarf)

- individuals and collective revelation as well as specific ethics in writing by people in specific geographies, people of color, queer folk, and
- as many of the above as possible to decenter argument and refocus attention on the combinatory apparatus of the reader and the writer through one of several non-instrumental approaches.

Environmentally, humans benefit from multiple approaches to poetry:

- expanding our understanding of our own humanity
- influencing social and political conditions
- clarifying the connections between/within the manifold,

all tactics associated with language, rhetoric, and prior poetry.

Arguing for only one approach engages with language in a way that can sharpen writing into an argument at the expense of improved accuracy across multiple scales. Functionally, environmental approaches to writing appear everywhere if we look at writers as organisms operating both singly and in groups within ecosystems. Adding this third component, ecosystem or surroundings, to the individual/society model of poetry/politics makes all these situations both easier to understand and more accurate. Environmental approaches do not exclude monoculture or things/people in themselves, since organisms are both aware of themselves as a whole and recognize (and often vociferously deny) that they are also composed of parts and connected in ecological ways. The simple necessity of saying "my hand hurts" appears to contradict the fact that things do not exist in themselves and are only performative. Without thinking of the body as the ecosystem of your pain, it's easy and fun to just notice that your hand hurts. Accepting that self, society, and surroundings all operate as ecosystems and together as an ecosystem reduces stresses of performance on all of them.

Groups Build Social Syntax

This chapter begins by questioning the self as a whole entity through works of Borges and Whitman. Group formations as a category build social syntax by scaling up and down, adding, and contractions. Group narratives build institutions and encourage people to value themselves through the group. Connectors between individuals and groups are recognized through Baudelaire's notion of correspondence. These various constructions of identity are built as a multiple where parts are connected and often scale in poetries like Donne's, Mallarme's, NY School, and Flarf. Grammaticalization appears as a scalable social and ecological method to build environmental consciousness.

* * *

A social syntax derives from group dynamics with the individual self partly identified with multiple groups and partly with the organism and its components. Jorge Luis Borges:

> There is no whole self. He who defines personal identity as the private possession of some depository of memories is mistaken. Whoever affirms such a thing is abusing the symbol that solidifies memory in the form of an enduring and tangible granary or warehouse, when memory is no more than the noun by which we imply that among the innumerable possible states of consciousness, many occur again in an imprecise way. Moreover, if

J. Sherry, *Selfie*, https://doi.org/10.1007/978-981-19-4870-1_10

> I root personality in remembrance, what claim of ownership can be made on the elapsed instants that, because they were quotidian or stale, did not stamp us with a lasting mark?[1]

Despite Borges cogent line of reasoning, individuals seek convincing narratives (sequences of events) about themselves to feel comfortable and to help them function. They then edit stories of those events that they continuously play with, filter, and sort. The combination of sequence selection and the skills to cope with elastic identity together construct a self that feels distinct and appropriate to a person. Acknowledging how I'm different moment to moment is frequently so threatening that I often reject what should be obvious: I function and consequently identify on several levels at once.

A similar process of identity creation occurs for groups and for the same purpose, that is, to build a functional model, not a perfect, but a defensible and workable site of group self. To build that group self, leaders seek cohesive vocabularies, order, and phrasing to accumulate into identities that the group, its subsets, and individual participants can apply to goals and external threats. Word choice and order in phrases sounds like a concern in poetry. "America, the beautiful" in one of our national songs reads quite similarly to the literal translation of the Chinese name for the US: "beautiful country" (美国 meiguo). And the official Chinese name is even more extravagant: (美利坚合众国 meilijian hezhongguo): The United States of Beauty, Advantage and Endurance. And all this beauty is subsumed under the linking adjective United.

Walt Whitman uses metaphor to identify the United States on two scales at once: "The United States themselves are essentially the greatest poem."[2] Many politicians continue to support Whitman's ideal of constructing a cohesive whole out of disparate parts, "states" that Whitman also equates with states of being as a psychological inter-intra-action. There are also poets who have exposed his well-documented, anti-Native racism as contradicting his claims of universality. Racism a good topic to address here as exposing Whitman implies that one action of the poet rules the value of his poetry and reputation through morality; his inconsistency invalidates his assertion. Shall I now reject Whitman entirely or can I appreciate his value but remind myself of his racism? Do I reject myself when I see my defects? Do we reject America for its flaws and move to an imaginary country pure of heart? The critique of his behavior implies that Whitman should have been perfect. But as June Jordan wrote in nuanced engagement with Whitman:

> I too am a descendant of Walt Whitman. And I am not by myself struggling to tell the truth about this history of so much land and so much blood, of so much that should be sacred and so much that has been desecrated and annihilated boastfully.[3]

Refocusing the group selfie where Whitman and I took our picture at the Grand Canyon, can I position how I resolve this conflict as a better way to understand myself? At the personal level, I make mistakes in behavior and judgment. If I forego the possibility of achieving perfection and accept that I have faults, have made mistakes, even big ones, yet continue to try to improve myself, poetry, and, by extension, the world around me, there, among multiple considerations, I'd like to find, rather I insist that I find, environmental methods and strategies.

Human error, writ through an organismic tendency to reduce labor, demonstrates similar dynamics. Industrial emissions of toxic waste and destruction of ecosystems raises questions about the value of our species in the biosphere. I want to question human desire to expand our range, assert control, realize independent agency, and uncontrolled reproduction as follows: What are ways to access the connectors that carry the species narrative so we can change it? Many critics from Jacques Derrida to Michel Foucault to Donna Haraway and onward speak to these connectors without thoroughly naming them.

How many different human narratives need changing? Is humanity's general narrative the only one that needs changing or are there subsets of humanity, like corporate leaders, with problematic stories about themselves? Since I support Borges description, each group must change their narrative. One human story doesn't speak to everyone but through rhetoric.

Each node of the network has its own point of view. Their narrative channels may be built into entire ecosystems, like deserts reinforced by protracted drought, like conceits in the extremes of care that increase sensitivity to insults, like identities built through communications media that require continual input from those media. Identities that protect us from our surroundings may be built into the very concept of an ecosystem as something we live inside of.

In the ecosystems of social media, news, public relations, and even traditional publishing, identities are formed by:

- repeating phrases associated with a topic or person
- capturing imaginations in frames with shared organizing principles like individual well-being, family, and anger
- institutions that people value and identify with as diverse as The Confederacy, the Federal Reserve, and *Poetry Magazine*
- the actual effects of climate change on individuals and groups.

Actual change agents like heat, work, and new tendencies in poetry:

- build new connectors to others and consequently more influence in the network
- function directly on climate like a mine owner deciding to dump tailings into local streams
- operate indirectly like essays through connections between poetry and ecology
- seem to have little immediate effect on the general public like tent cities built by income inequality and job loss due to technology, but the dissatisfaction they create drives social change.

How can we intrude on the capitalist story of growth and expansion as it continues to drive pollution into our surroundings, continued corruption of social structure, and continued anxiety, fear, and frustration for individuals working to transition this tale we've created for ourselves to a more balanced one? One way is to recognize that Capitalism actually created what Euro-American culture understands as environment. Nature is not just the source for capitalist production ("webs" and "sinks"), but as Moore says:

> It is the field upon which capitalism unfolds. And we can go still further. Nature is no static field, but is itself renewing and evolving in cyclical and cumulative fashion. Nature is, above all, historical.[4]

While surroundings are part of social relationships, surroundings are also implicated in the environmental sense of syntax, since syntactical order develops in ways partly determined by habitat, in other words, the concepts surrounding the sentence and adjacent sentences. Moore again:

> First, capitalism does not 'produce' nature in a linear fashion, but is an evolving whole that joins the accumulation of capital, the pursuit of power and the co-production of nature. Second, capitalism is not a structurally invariant, monolithic Society, acting upon a structurally invariant, external Nature. Rather the history of capitalism is one of successive *historical natures*, which are both producers *and* products of capitalist development.[5]

While our surroundings are implicated in social relationships and social relationships form our surroundings, habitat's connections to ourselves need to be continually renewed since the organismic binary, the dictatorial component of self, obscures the location of our selfie as anxiety and real threats increase. Our faces get larger in the frame and we no longer recognize where the picture was taken. Social threats drive changes in group formations, too.

On the social scale of change, group syntax evolves like individual genomes with inheritance and epigenetic correspondences such as the history of US race relations and the murders that swell the ranks of Black Lives Matter. As culture workers and citizens together, we can affect social evolution and consequently climate change, but we need to accept the heterogeneous characteristics of any change group. Insisting on homogeneity in the group at all scales keeps the group from evolving toward group goals because any complex group with complex goals needs to fill multiple roles.

Among those roles will be those who operate most effectively within the group and those who link the group to other groups. Wilson:

> Does benefiting the group require overt altruism on the part of individuals? If not, then group selection can favor mechanisms that organize groups into adaptive units without strong selection against these mechanisms within groups.[6]

Within the poetry group of language writers, there were those who focused on building group identity, those who linked the group to the rest of the poetry world, those with strong affiliations to the group, those who had weak affiliations to the group and were also affiliated to other groups—a heterogeneous set of roles requiring more than one type of writer, more than one style of writing. In addition, most individuals played multiple roles with a few acting only as poets, only as publishers, only as teachers, or only as organizers of public events.

In that way, the language writing group was able to evolve with multiple perspectives on its common goal. These diverse points of view were then able to stepwise occupy multiple niches within the larger world of poetry writing, academia, publishing, and attracting the public eye. The ambition inherent in the historicism of language writing also made language writers write in many genres including pastiche, homolinguistic translation, lyric poetry, field poetry, list poetry, minimalist poetry, academic essays, popular essays, personal essays, essays on poetry with political overtones, and memoir.

As the language writing group evolved, it influenced different scales of social structure from individual poems to archives to generating other groups of poets both in reaction and to clarify details of language writing strategies. Social identities within groups like poetic genre, organization building, and distribution networks scale from brain to social group to habitat. Analysis of details and fragments in biology, sociology, and poetry does not make them whole by virtue of environmentalism but can be shown to have characteristic, similar, multiple, and diverse structures, similar not identical. Individuals retain their differences at all levels in a common landscape (map) while some concepts and processes scale from the cellular level to the organism to the ecosystem and even to the planetary. This scaling of poetic components operated as a network that scales like a genome. Hyunju Kim et al.:

> To characterize planetary-scale biochemistry, we constructed biochemical networks using a global database of 28,146 annotated genomes and metagenomes and 8658 cataloged biochemical reactions. We uncover scaling laws governing biochemical diversity and network structure shared across levels of organization from individuals to ecosystems, to the biosphere as a whole. Comparing real biochemical reaction networks to random reaction networks reveals that the observed biological scaling is not a product of chemistry alone but instead emerges due to the particular structure of selected reactions commonly participating in living processes.[7]

Universal scaling shows how the connections between different levels of the biosphere make it possible to imagine a culture that highlights scalability and similarity with personal and group differences to encourage changes in the processes of the six critical industrial processes.

Universal scaling does not mean that everything scales or that scaling is smooth. Limits to universal scaling appear overtly in the organismic binary, because people can't easily encompass the entirety of the planetary change that industry has caused. One of the reasons for failing to understand climate change and act on it are what Lynn Keller and others call "scalar discontinuities."

> The scale of human attention tends to be small and self-centered, not attuned to what Rob Nixon has called 'slow violence.' Even changes that geologists easily recognize as alarmingly rapid can, because of their geographical distance and their suprahuman scale, remain abstracted from ordinary lives.[8]

Keller cites Spahr to show the limits of scalability in a state of environmental panic:

Unnamed Dragonfly Species
They were anxious and they were
paralyzed by the largeness and the
connectedness of systems, a large-
ness of relation that they liked to
think about and often celebrated
but now seemed unbearably tragic.[9]

That "attention tends to be small and self-centered" certainly results from the organismic binary, intimacies of poetry, and the limits of perception as well as cultural analogs at other scales. This fact of biology and social interaction is widely appreciated. Yet, it is clear for both Keller and Spahr that the inability of some human imaginations to scale up to the problems of climate change shows differences in perception among individuals rather than a failure of important frames of the biosphere to scale up and down. Keller's seems to me to want the species concept to scale and to adhere to assigning blame through the term Anthropocene.

The problem for humanity and the biosphere as a whole is not the species which doesn't scale, but the assemblages built by scientists and engineers, funded by corporate leadership seeking profit and hegemony.

These assemblages—combinations of human, chemical, and constructed entities—operate as polluters in all six industries that cause climate change. What scales is the ambition of corporations seeking power and profit. What scales from the microbiome to the whale is the ambition of all species to reproduce by altering their surroundings while at the same time adapting to conditions that the species cannot change.

The concept of species does not scale and must be viewed relationally. A species does not scale because its needs stem from homogeneity while the next scale upward, the ecosystem, produces value from diversity. So, the term Anthropocene misconceives the causes of climate change as a species error and presents yet another error of moral judgments. The solution lies in changing the goals of corporations and the characteristics of their assemblages which require social change and cultural change away from blame toward support for solutions.

Let's forget the Anthropocene for a bit. Where can environmentalists put their attention to be most effective as an abstraction and as a practice? As Jacques Grinevald says in his introduction to Vernadsky's *The Biosphere*:

> The functional concept of ecosystem...has no geographical boundary outside the observer's choice. Its extent is defined by the scale of observation.[10]

Grinevald's dynamic boundary is another framing of the reason I originally titled this book *Selfie*. If poets choose the same scale of self-expression and single group identity, poetry's effects on society and the biosphere will continue to dwindle. Derek Woods points out:

> The scale-critical subject of the Anthropocene is not 'our species' but the sum of terraforming assemblages composed of humans, nonhuman species, and technics.[11]

Woods "terraforming assemblages" (a synecdochic synonym for Barad/Bohr's "apparatus") supports but is not identical to the idea composite agency. Woods refers to a combined entity while composite agency refers to agential effectiveness.

Even in an environmental model, second-order systems cannot be separated from poetry. Scale significantly depends on the observer because both Keller and Spahr can also imagine that poetry operates in combination with components that scale up and down even as some minds have difficulty, even resistance, grasping the scalar changes. Keller even points to Spahr's use of punctuation that scaling occurs:

> Spahr's not putting punctuation between each species name and the sentence that follows invites readers to think of the two strands of the work as part of one thing.[12]

The "scale critique" in Keller, Woods, and perhaps Spahr attempts to trouble ideas of transparent, smooth scaling, an easy target. But for culture to be effective in changing ecosystems, finding the components of culture that scale up and identifying similarity at different scales needs to trigger inferential identification and improve the chance to change people's minds about difference and the value of unique acts and creation.

Shame and fear seem less effective than a renewed view of the biosphere, societies, and selves. A workable model of scale helps to understand what aspects of poetic composition, reading, and interpretation support climate change remediation. Scalar discontinuities support difference, componentization, and undermine the all-in-one indissolubility of the universe in primitive holism and simplifications in religions and eco-piety. Scalar properties are then useful to highlight in order to avoid homogenizing humanity, as Keller fears, where difference and contrast seem more useful when connected to difference in a non-evaluative hierarchy like taxonomy. Poetry and prose.

Many people easily grasp that humanity is in trouble because industrial pollution is rendering multiple habitats increasingly unfriendly to existing species. Further, the inference machine of the mind also tends to scale; humans associate widely. But not all components scale in the same way because as Woods notes, "the observation and the operation of systems are subject to different constraints at different scales due to real discontinuities."[13] Similar and different, environment does not operate as a whole, but many processes do act as a whole in spite of their component construction. Environmental connectivity and the physical world scale while perception of them often prevents people from understanding what is bigger and smaller. Further, some people cling to their assumptions, refusing the scales that they fear will compare them while they want to be at liberty. People experience Keller's "scalar dissonance" even as scientists increasingly uncover scaling properties in the biosphere.

As Timothy Clark puts it, "Non-cartographic concepts of scale are not a smooth zooming in and out but involve jumps and discontinuities."[14]

Some universal scaling, some partial scaling, some scale discontinuities, and some non-scaling are all features of the biosphere. Importantly, many organismic functions and perceptions don't scale. In connecting the reader to concepts from the biosphere, successful operational change proves which relevant components scale, since that is the connection traveled as individuals start to think differently about themselves in spite of those non-scaling views from the perspective of organisms.

As a connector myself, I want to know where language might be able to scale. Particular social and biological syntaxes govern development and progress at multiple scales. Syntax operates even in processes that don't scale. Children crawl before they walk. Building houses precedes the need for a fire brigade. Plants precede the animals that eat them. Reproduction by cellular division precedes the greater genetic safety of bi-parental sex. People make marks before they read but read or at least listen to poetry before writing it. These kinds of social syntax connect over time populating a socio-environmental template. Syntax evolves as in Bernat Corominas-Murtra et al.:

> The first stage is the so-called Babbling, where only single phonemes or short combinations of them are present. This stage is followed by the Lexical spurt, a sudden lexical explosion where the child begins to produce a large amount of isolated words. Such stage is rapidly overcome by the two words stage, where short sentences of two words are produced. In this period, we do not observe the presence of functional items nor inflectional morphology. Later, close to the two-years age, we can observe the syntactic spurt, where more-than-two word sentences are produced.[15]

In one possible scenario, I summarize from Luc Steels, building social interaction using human language starts with the development of both increased mental capacities and vocalic apparatus. But does external development of efficient gait and opposable thumb precede mental and vocal abilities? The point is that evolutionary thinking always operates through a syntax, even multiple simultaneous ordering, one word, thought, action, and morphological development after another. Then human language structures begin to exceed other primates.[16]

Other species, too, have sophisticated modes of communication and methods sometimes different than people. But like humans, they perform their identity partially in languages without words, using syntax and other methods related to ordering events. Monica A. Gorzelak: Trees feed and communicate through root development aided by fungi.[17] Gibbons and mice sing. Bats and platypuses echolocate. Bee hives network in complex ways that prioritize specific behavior for each class of individuals.

Social syntax orders phrases and diction passing between people in speech and writing. At other scales, it orders how people think about themselves, their ideas about the world, embodied actions and group actions. While not all cultural development has such clear and physical correspondence as in human language, both society and the biosphere progress through similar sequences. Evolution is such a narrative syntax at another scale. Writers use methods such as metaphor and synecdoche to demonstrate relationships to those whose behavior needs change. Poetry uses scalable tactics and strategies.

Even when apparent connections mislead us to consider different processes as similar through unverifiable superstition or different species as related through similar processes as in convergent evolution like bats and birds, these parallels condition the behavior of individuals and groups. They prepare the ingredients and add them at the right time. They taste and correct the seasoning. Because of the complexity of interactions, society may end up with the tasteless, efficient soup of corporate happy faces in the same meal as a spicy, murderous meat of visceral reactions to perceived repression of freedom. We can't ignore the rhetorical tropes in the stories we tell and how we link them to choose among possible actions. Isolating metaphors produces species extinction.

On the scale of reading and writing poetry, evolving connections that Baudelaire called "correspondences" "regard" poets as they perform and link their writerly and social identities. Correspondences feedback to each individual writer from their social milieu, and habitat is unmistakable but difficult to parse through one-to-one correspondence alone. The links between individuals and

- other poets in the poetry world for writing
- family, friends, and colleagues for daily life
- significant issues of the day for political action, poetic persuasion, and satisfying poets' aspirations and needs
- the biosphere for sustenance, knowledge, inspiration, beauty, and reassurance

build phenomena composed of multiple entities with unique names, often throughout the manifold of interactions. Social syntax accommodates difference through these many-to-many relationships.

An example of theme and form: Correspondences derive from literary, listy, spacey, speechy, and procedural syntax. *Literary (verb-oriented) syntax focuses on complex psychological interactions as in Donne's "The Canonization". Listy (noun-oriented) syntax helps understand social structure as in François Villon's "Testament" and the second book of the* Iliad. *Spacey (duration-oriented) syntax shows relations between ideas and ecosystems especially where space acts as a measure of time such as Mallarmé's "Un Coup de Des…" Speechy (personal) poems, as in O'Hara and Ted Berrigan, focus on daily language interaction but can also spoof philosophy in an American way. Syntax determined by procedures (socially conceived) establishes political positions in Dada, Oulipo, Fluxus, and Flarf, as in Nada Gordon's* Folly.

Correspondences between these different strategies of poetry and conditions of habitat help individuals decide on the best, most useful, entertaining, correct, and effective order for their poems, for their behavior, and for their society through their book buying, voting, and other social support. Can syntax help decide how to engage with our surroundings? *Types of sequencing correspond to what we see happening at other scales. The syntax of rainfall, seasons, and lives shows how longer periods of time, longer than instants of awareness, work. They build multidirectional connectors between writing practices and events. We see it predicted by the sentence structure, informed by brain structure, and, as in this extended paragraph, in an altered, relative positioning, font, and spacing of sentences.*

Phrase, sentence, and paragraph syntax act as models by mirroring the self in a way that people desire and by building further connections that scale up to ecosystems. *Poems look and act like us. Poems, bodies, and group identities are built by performances through those connectors that create correspondence, encouraging us to value the model.*

Since individual identity corresponds to larger frames like group identity and habitat, modeling correspondences between individual and social connectivity on the biosphere constructs a flexible, sustainable culture. *Solutions from biomimicry (modeling social activities like growing food and generating energy on biological entities and processes) have been vetted sometimes for thousands of years in agriculture, clothing, and construction.*

Rather than solutions focused solely on competitive acquisition of limited resources, manipulation of profuse resources, or moral avoidance of compromising interaction, diverse views of resources strengthen society. But it isn't sufficient to describe these connections. Pastoral poetry has been describing nature for 2500 years. Alexander Von Humboldt's *Cosmos*, volume 2 is composed of a 400-page list of pastoral poetry through the mid-nineteenth century.[18] Draw explicit connections and follow them.

Changes in poetry look like social and habitat change because similar dynamics of change propagate throughout the biosphere; a change in one increases the likelihood of a change in another. Unfortunately, science's precision has conditioned Western culture to expect even non-linear systems to exhibit single value, thing-like solutions. If our cultures took into account the differences between integers that act like organisms and the rangy probability of the biosphere, it would be easier to protect civilization from the worst devastations of climate change. Leadership's expectations of quarterly, continual group would change.

Scaling down: Contractions reduce the energy required to write and speak by condensing the semantic value of two words into one—a kind of efficiency that changes meaning. Reduced reliance on authority changes *let us* that asks permission to *let's*, a collective act. Some authorities still think of contraction as an affront in an effort to retain influence. Small-scale alterations in grammar produce large-scale resistance. Instead of appealing to authority to allow you to act, you simply call on the group of equals to use their coherence to act together in broader social cohesion. Reducing labor through contractions actually supports conservation of energy. Contractions bring people as well as words together. A collective resource replaces the evaluative hierarchy of a single correct grammar power structure.

But not all Grammaticalization increases efficiency. The phrase "I am going to" instead of "I will" implies a loosening of individual power, uncertainty about outcomes, time wasted with more words, and goals delayed, a threat to the progress of capital insofar as telos identifies a political perspective. Grammarians, in naming this process Grammaticalization, align themselves with the evaluative hierarchy.

Yet, would you be willing to forego contractions if you could require mining operations to clean up their tailings? Would you be willing to forego your right to assault weapons to save the lives of children you do not know? Complete, consistent, scalable environmental culture isn't a requirement for reducing emissions, but it helps people comprehend their connectivity. Preserving rights of access to ecosystem resources, even social ecosystems, makes climate change remediation that much more difficult. What if your transportation costs increased 50% as electric cars and trucks became more popular? Existing biases, habits, individual expectations, and vested interests resist those changes together and are altogether difficult to overcome.

The renewed focus on particles of language overcomes the semantic poverty of the *minor* entries of the lexicon as opposed to sexy top predator nouns. Semantic values increase, too, when we add new words, such as cyberspace and webinar, to represent new connections that are built by technological networks, and matcha that brings a global perspective to our tea drinking, and Brexit that defines a nationalist response to globalization, and flexitarian that specifies dietary awareness about the industrial food chain. These words and the material entities they point to build connections from individuals to their societies and habitats. New meanings arise from tighter cultural linkage to our surroundings built by ecological networks. Technology, supply chains, communications, and personal proclivities establish meanings that are reinforced by the immense surplus of extractive industries and governance. New language around production helps accept new methods of manufacturing as in Biomimicry.

The incentive for leadership to use environmentalism for social control is revealed by how much corporate powers benefit even when creating surplus and channeling that oversupply to those resisting their extractive processes. Individual and group resistance to the norms of habitat destruction provide additional energy to corporations in establishing a culture of continual revolution, further blunting the effectiveness of resistance. Vaclav Smil: Among the complexities that arise from perspectives on evolutionarily stable strategies, readers might also begin to question their own assumptions about the fixed value of both sustainable processes and monoculture.[19]

Some of the clichés of the environmental movement, such as sustainability, need to diffuse more widely and to specific flora and fauna of meanings to be effective as environmental concepts. I mean no criticism of the term sustainability, but rather a scaling of the idea throughout culture, even though we risk some conservative framing such as inflating the value of tradition, legacy, and the litany of language structures that support aristocratic and elitist politics. Sustainability has a range of meanings outside of ecology. For entire texts and named groups of texts, it might refer to how they are taught, thought about, and respected by forward-looking writers, so they can continue to be read.

For intra-textual continuity, I'd like to find ways to keep the reader reading without compromising a thorough set of references to events that are obviously connected to the text as well as opening new channels of connection. For example, reducing the number of placeholders in syntax increases their semantic value in sentences, a practice intended by most of the poetry that I have read, edited, and performed by the equals signs between each letter associated with language writing. I would call this increase in semantic value parataxis but the connectors of hypotaxis are hardly placeholders. Rather these connectors imply a more integrated, less fragmented meaning/world view as in the rolling sentences of Marcel Proust and Machiavelli. I more likely refer to intra-textual edits such as reducing the number of adjectives that telegraph meaning rather than letting it emerge from the syntax.

Equally, the phrase "ecopoetics," which has surfaced in the discussion about poetry related to climate change, makes it easy to discuss a type of nature poetry as a unit but it also makes it easy to gloss over the poems themselves. The commodification by the abbreviation "eco" of ecopoetics both attracts attention and limits readers' responsibility to engage the actual work and to question how this kind of nature poetry is different than prior pastoral writing. Named schools of poetry including language writing, Surrealism, and Imagism act as contractions of the individual poems and poets gathering adherents even as they mask individual writers and works. Group formations are often confused with oppression, even though every organizing category, including groups, reduces the complexity and options for making mistakes. At the same time, group formations make it more difficult to find new connections between components of the biosphere, thought, and poetic phrasing. Group formations can also act as agents of oppression.

Reusing inaccurate placeholders like the label academic poetry that refers to a level of abstraction that substitutes mental gymnastics for personal expression, not poetry by people in universities, commodifies and glosses over details about the different poetries written under that label, clear-cutting in a forest and redlining of poor economic risks among city neighborhoods decreases the value of each. But if reducing the use of intermediate groups like ecopoets only highlights the value of each individual organism, poem, and prerogative, then control of society reverts to corporate oligarchs that play on our tendency to identify with and protect our organisms. There are many pitfalls so I suggest that our best bet is to attempt to increase the accuracy and reach of these intermediate institutions, intermediate terms, and connections, even problematic ones, rather than oversimplifying social syntax with over-reaching ideas like gut intelligence or essentializing poetry as sound.

Adding surroundings to the binary of individual and society, as opposed to subtracting the group from the individual, improves our chances of avoiding these traps and removes some of the pollution of singularity. A selfie that is only a picture of me and not who I'm with and where I am reinforces that narcissism that undermines the more complex world we acknowledge with environmentalism. Environmental poetry uses what I know about me to connect to others and include my surroundings, your reading, and our connections in common habitat. Simply relying on me is not a sustainable strategy and plays into the clutches of corporate control.

Some Structures of Syntax: Parataxis & Hypotaxis

In this chapter, I reflect on two syntactical structures through the work of Hejinian and Silliman. Each syntax emphasizes different interior, transitional, interactive, and exterior connections. Parataxis values independence of individuals that emphasizes the general idea of connection between things without specifying the type of connection and encourages reading the spaces/connectors between sentences as material. Hypotaxis emphasizes interior connections and specifies an individual's characteristics and the writing process. Solutions to hard problems like climate change are not easily uncovered in syntax of sentences alone. Together these syntaxes combine as a method for ordering words and creating flows toward cultural action on climate.

* * *

Bob Perelman's notion of parataxis emphasized language writers' focus on creating a social poetry based on technique in the 1970s and 80s:

> Parataxis is a dominant mode of post-industrial experience. It is difficult to escape from the atomized subject areas, projects and errands into longer, connected stretches of subjectively meaningful narrative—not to mention life. As objects of the media, we are inundated by intense, continual bursts of narrative—twenty seconds of heart-jerk in a life insurance ad,

J. Sherry, *Selfie*, https://doi.org/10.1007/978-981-19-4870-1_11

> blockbuster mini-series ten nights long—but these are tightly managed miniatures set paratactically against the conglomerate background that produces them.[1]

Since then and in response, many younger writers have reinvigorated narrative in spite of the continual fragmentation of attention by social media in the twenty-first century. From the environmental perspective, the components of language, individual identity, society, and biology operate both independently and together through multiple syntactical structures. They function independently and have separate names, but like organisms, they do not thrive alone. Writing with only one kind of syntax as in Silliman's parataxis serves a didactic purpose that may be needed when one syntax seems overworked and for stylistic consistency. But one kind of syntax cannot include all the varieties of and requirements for writing. Using different sentence structures—parataxis and hypotaxis—over time and all in one place serve to reestablish balance in an autopoiesis of style that cannot permanently exclude other constructions.

Financial interests have achieved too much power in recent decades. In response, writers have moved toward sociocentric approaches by developing political language, paratactic structures that reflect political conditions in culture, and writing about environmental dynamics. One-way processes can be fostered is at the level of syntax.

Conventional flows continue
to dominate town crying.
Corporations corrupt governments
with pollution by lobby.
Individual desire (as opposed to need)
imitates privilege for majorities
in all major nations and cultures.

While one has to be cautious in making generalizations about groups, the groups themselves are generalizations in that the individuals in them choose to select (a) component(s) of their identity to focus on in the group. Gender as difference operates in specific ways for the transsexual community, but not the same ways for all individuals. It's definitional if I categorize it, even though the gender is trans and fluid. As Trish Salah speaking as Tiresias classifies in her book *Lyric Sexology*, she was not in those categories but was in these categories:

> This is the introduction to this book, my introduction, my lyrical sexology.... I am not a transsexual. Or an intersexual, or a hermaphrodite. (Hermaphroditus can write her own damn book.) I am not any of those things you have words for now. You don't have words for what I am. What I was was this:
> I was a dude.
> Then I was a chick.
> Then I was a dude again.
> Hah. You didn't think we said 'dude' or 'chick' in what you call ancient Greece, Hellenes, etc. Think again.[2]

The unity of these differences that is two sides in one difference brace readers for thinking in categories. In multiples rather than binaries, categorical understanding turns out to be critical for environmental thinking, too, in the sense of taxonomy and as an alternative kind of hierarchy to evaluation. While there are subcategories among transsexual communities, such as the short-lived "Sexy Black Transsexuals" Facebook group, color is not the center of the identity of transsexuals taken as a whole. Gender and transsexuality have a different kind of connection than color and transsexuality. Again, this is true for the group category but not for all individuals. Yet, gender scales from individuals to social groups, and the category of gender is largely socially determined.

Alliances, a kind of political connector, play an important role in political and environmental power, so individuals in a political group representing white progressives might well ally themselves with a Hispanic group to form a larger group. Class plays a complex role in any community stressed by oligarchic and financial forces through constant assertion of leaders' rights and denial of privilege to those not in the top income layers. Groups stressed by lack of access to resources—the poor, people with disabilities, youth, and others mentioned already—generate the counter currents in society. These groups may well ally with white progressive/Hispanic groups. We see among the GOP in 2020 an alliance between financial sector conservatives, fossil fuel conservatives, evangelical conservatives, and libertarian conservatives. These alliances are fluid in the sense that their connectors can be detached and reattached. We saw Democratic political majorities in Southern states in the US fade as Democratic connections to labor were replaced by GOP connections to conservative groups that had grown up and allied to form new majorities.

Then groups of groups connect to form larger social structures like political parties, nations, societies, and diverse communities like the poetry world, where several, many, different styles of writing coexist in ecosystems of greater and lesser harmony. While we can talk about groups forming syntactically, this complex set of relationships makes it clear that we need more than one kind of syntax to explain how diverse groups operate together.

Sentences can be organized in both paratactic and hypotactic syntaxes. The paratactic is political in that it stresses both organismic independence and assumes relationships without specifying them. In resistance to humanist sentimentality parataxis levels the field. In Silliman's paratactic sentences, the units appear as parallel and similar even when their importance varies for each reader.

> Stubborn as a mule, sir, stubborn as a mule. Log fort. How then can the sense and the truth or the truth and the sense of sentences collapse together. As map could expand beyond the margin.[3]

Silliman's parataxis sponsors naïve equality of sentences and in the borderlands of the connections between them. Yet, each sentence and the spaces between them differ. The sentences like individuals in a weakly or democratically connected society appear to stand alone, simply juxtaposed. He does not differentiate the types of connections transparently, simply presents them in an ecosystem of arboreal sentences, but does focus on groups of sentences using methods like the Fibonacci sequence to determine the number of sentences grouped in a paragraph. Parataxis appears as an idealization of social relations: equal access to readers for different meanings.

Lyn Hejinian's hypotactic sentences specify connections. They drive meaning through a variety of complex, biographical events to support, torque, and rein in the thrust. They accommodate and represent change within the sentences compared to the way Silliman represents change between sentences. Between sentences, Hejinian establishes greater continuity with thematic and often logical links. Her hypotaxis reveals the relationships that are difficult to talk about or hard to see in a world of digitized parataxis.

> Indeed, in these times, it seems, we back away from thought altogether, we scarcely think at all, given the diversions or the mechanical aids that block thought by making it unattractive or unnecessary.[4]

Hejinian's sentence about thought is constructed tentatively, hypotactically, maybe even run-on. The connections between words and phrases in the sentence lead through the thought. The thought builds through the sentence feinting forward so that the reader is drawn toward agreement.

Parataxis focuses on sentences as units with uncertain relationships between them, representing political independence. Hypotaxis looks through a sentence connected in a specific way to the rhetoric of self, society, and habitat. Either one alone is insufficient for the environmental model to show the complete sets of organisms and relations within and between individuals, societies, and ecosystems. Organisms appear alone, but connectors between them are essential to their survival. In units such as voter coalitions each person has but one vote: parataxis. When individuals want influence, they call each other, knock on doors, and put opinions in media. Depending how much time and effort these individuals want to apply to interests like applying for government contracts, they can develop policy statements and links from their individual needs to political actions. They can influence voter coalitions by building specific connections to their interests: hypotaxis.

Between identity groups formed around poetry, parataxis reigns through mostly separate and juxtaposed differences in style and intent. Language writers, Flarf writers, New Narrative writers, people relating to writing not only through style such as Black writers, Hispanic writers, gay writers, lesbian writers—the list is long and society complex. These groups have an independent identity but are linked by various usages of language, schools of thought, readings, publishing, and distribution. Environmentalism asserts that the relationship of specific texts to the group is isomorphic. Some aspects are similar, some are different, and the subject is polymorphous. I, for example, sometimes relate to people as an editor, sometimes as a writer, sometimes as an American citizen speaking for freedom of expression, and simultaneously for taking responsibility for my own way of framing ideas. I am not fixed, and subjectivity itself must be questioned for people and their cultures to avoid destroying their surroundings.

Language writers link their techniques and intentions to politics but still call themselves poets, not political activists. Their subjectivity is multiple. Black Arts Movement writers of the 60s and 70s linked diction and syntax to the language that Black citizens use in resisting marginalization and repression, but also individuating the Black experience through hypotactic emotions. In a more recent expression of Black culture,

Claudia Rankine writes: "Some years there exists a wanting to escape—"[5] Chinese "Original" poets, like Che Qianzi and the early work of Huang Fan, link their style to the Chinese character and the etymology of the specific character in a way that is both similar and different from Ernest Fenollosa in *The Chinese Written Character as a Medium for Poetry*.

Translations of poetry have so many possible vectors that Sun Dong and I created *The Reciprocal Translation Project* where multiple subjectivities are considered together to more fully reflect the "identity" of the original poem than any single version.[6] In *The Reciprocal Translation Project*, six American and six Chinese writers translate each other. The book combines literal, lexical, etymological, and contextual methods of translation to engage new possibilities for transcultural and trans-linguistic literature. Each poem was first literally translated and then poetically translated by three poets of the other language. Instead of a single translation, there is one literal translation and three poetic translations for each piece, acting in concert and conversation. Che Qianzi:

When I write a poem
I think of you
Feel there is some coincidence
I want to write a poem
But do not know what will happen later[7]

Within identity groups formed around poetry, parataxis operates in the way each poet writes to distinguish their work from the group in a parallel method. Both Ashbery and Silliman wanted to deny the respective groups that spawned their visibility as they developed identifiable, individual careers. Yet, beyond the classroom and the exclusionary practices of social groups, parataxis is rarely achieved since the main form of influence within the group is not the binary but rather the hypotactic influence of style and intent, complex interactions related to the readers they attracted, and to a specified canon of past poetry. Each of the poets in the group is separated by a period; all of the poets are linked either strongly or weakly, continuously or intermittently by group influences that can affect the individual or the group and in both directions. The individual writer emerges from the group of influences of their society and their reading. The network does not undermine individual creativity but allies it with the poet's network of influences.

In the collective editing that forms the social network of many publications, giving each editor a veto emphasizes the paratactic, but before the veto is declared, hypotactic influence is likely asserted through speaking about the choices of which poems to select, themes of an issue, and what changes to suggest to the author. Giving the majority its choice includes both hypotactic and paratactic influence. Subverting that process of choice isolates hypotaxis. These processes also operate for states and ecosystems.

Environmentally, each dimension of the biosphere from intracellular chemistry to entire ecosystems supports different decisions about how to remediate it and the forces that affect it. Each dimension also opens subsequent dimensions of decisions in a way that is like decision-making for writers. When a new species is introduced into an ecosystem or expands beyond its established proportion of resource utilization, everything in the ecosystem adapts. Sometimes that alteration changes biases for a long time. As I say in *Oops!* "snowball earth" changed respiration for the entire planet when algae blooms produced an abundance of oxygen.[8] Sometimes the new species or increased population is absorbed within prior processes like the regrowth of forests previously cleared to farm. In the isomorphic relationship established between writing and ecosystems, it's easy to see how writing and editing are related as well as different. Reading and writing are also famously connected in specific ways. Making distinctions this way reveals how an ecosystem such as an organism, a personality, or a piece of writing functions syntactically in sequence rather than setting up a fixed hierarchy like good, better, and best writing that separates writing and people by job, social class, and value judgment.

While paratactic juxtaposition will not save the planet, combining without the bridge of gerunds and other grammatical links implies intra- and intersubjective environmental identity while not identifying specific ecological linkages like the nitrogen cycle and the social process of getting poetry published. Parataxis increases the amount of information to digest and the range of possible meanings. As the connections between sentences in parataxis move toward randomness in their semantic leaps, the amount of information becomes indigestible and meaning begins to be generated more by the mind of the reader than presented on the page. These are volumes of data operating in ranges, not the votes of value judgment.

The characteristics of paratactic joinery appear also in:

- lists
- bi-parental reproduction
- invasive species
- grafting of plants
- GMOs taking over a field
- libertarian politics
- alliances such as the Treaty of Westphalia that established the validity and inviolable independence of self-determining states
- political campaigns where claims of connection are less important than juxtaposition
- advertising where beauty is juxtaposed to products and
- other ways of letting readers infer connections that the writer might be unwilling or unable to make overtly, factually, or logically.

Parataxis opens new possibilities with concomitant risks and changes of focus. One is less than sure of the outcome of the paratactic connection; the trial and error of natural selection are invoked in another link between language writing and ecology.

Writers love these paratactics because it's relatively easy for them to conjure up the fuller syntax of a thought. Parataxis is a low energy solution. Readers can build connections between fragments, but many readers cannot get beyond a partial understanding and feel confused by having too many options. These careful, shall we call them anxious, readers prefer instead the personal affirmation of narrative and hypotaxis with more connections clearly specified. On the other side of the coin, paratactic comic books sell in high volume and lists of things sell well as fact books. Even at a poetry reading, a good list draws applause as the audience establishes connections similar to that of writers and feels exercised. The breadth of the field being measured changes this view of popularity since literary publishing, commercial publishing, and list-writing operate on audiences of different scales.

Detailed hypotactic connections confirm willful change and assert authorial control. They suggest that we know something specific about how one is connected to another. Intuitive writing practices reinforce the reader's self-image by drawing specific, well-understood connections. Sophisticated writers draw connections that readers see in the details of their lives and minds. Hypotaxis is well-understood through:

- political campaigns that appeal to specific needs of identity groups such as Trump's appeal to those Americans left behind by neoliberal globalization,
- trade (Adam Smith and globalists),
- supply chains (Fred Moten & Stefano Harney),[9]
- corporate takeovers[10]
- genetic modification where specific linkages are placed for "improvement" of species. (Methods for inserting specific genes in DNA is spoken about as editing and I would suggest how metaphor is physical and material.)

Identity poetics, as American poets pose it, questions whether there is such a thing as a revolutionary subject. It presents sequences and sets of relationships within a group, an ecology of selves under conditions of oppression. Connections are hypotactic but often omitted and assumed to be known and understood by the reader.

Questions about the revolutionary individual are related to capitalist, racist, and sexist dominations that are revealed by working together in Mao Zedong and Antonio Gramsci and through social demand in Ernesto Laclau and Alan Badiou. Barad and other New Materialists claim the subject/object (apparatus/object) binary is fused because things don't exist alone but in phenomena for specific uses and measurements.

Environmental poetics maintains that things and people cannot exist alone and operate in sequences that can be referred to as syntax and process at different scales. Further individuals understand themselves to present in categories and to function as groups. The perception of myself as an individual is momentary, but it lingers adorning the page. Poems don't exist by themselves. They link to the author, to prior poems, and to poetry as a category. They link to the reader who activates the poem by reading. Otherwise, the poem sits patiently on the page as a series of black shapes on a white rectangle. In some ways, the poem is always being read. The meaning of a poem is ecologically determined in the sense of where and when it is read with a push in certain directions from the author via an assertive or vague intention. Some authors hope the changes they introduce will be inherited by the next generation of writers, an aspiration for writing to introduce epigenetic change.

Capitalism and the violence of the state create a subject concentrated on labor. The subject's interests and conditions of life become more and more aware as subjects (note the plural) work together and see each other experiencing the pain of exploitation. The condition of subjectivity must be plural for activation, although, to paraphrase Hejinian, this *thought is [also] the bride of* [some situational] *thinking* about oneself in the context of being with others. Subjectivity in environmental terms is as plural as the pronoun "they" is singular. Exploding the binary exposes the revolutionary status of environmentalism vis à vis humanism even as environmentalism accepts much current social structure as inherent to human biological and social syntax, acting under those terms through evolution.

These ecological conditions play out in both identifying a self-image and group formations. Have poets in their individual fervor ignored the evolution of the connections between them that constitute the forms of their poetry as a society? The twentieth-century focus on language would suggest that is not the case; poets know they are similar even if they resent it.

Further, poetry is subdivided into groups where poets read members of the group differently than those outside the network and with curiosity about those on the margins. And each poem that each poet writes is different with different connections to potentially different if connected readers. I have spent a lot of time with Velimir Khlebnikov's super sagas but not read much else of his work. I have spent a lot of time reading Ashbery's early writing but only recently focused on *Flowchart* and *Girls on the Run* where a new form of concerted environmentalism appears:

Why not? Why not just give yourself, one time,
to the floods of human resources that are our day?[11]

The social structure, the militarism of reading certain poems, of reading some poems more closely, and of reading poems for different purposes ranging from enjoyment, to learning, to career enhancement gets so complex that it is easy to forget the community of connections of reader, distributor, publisher, editor, and writer, replacing it with the easy contrast of the organismic binary. All line up in neat columns and phalanxes falling apart as those human resources fail to meet expectations.

Syntax, whether grammatically populated or juxtapositional, allies with punctuation to make its joinery more comprehensible. A space between words or paragraphs can be another coupling as especially in Silliman and the field poems of McCaffery. Specifying these connections looks like received form with the same strengths and weaknesses but can also establish new ones. The variety of connections, some strong some weak, some bidirectional some one-way, some continuous some intermittent, corresponds to social cohesion among groups.

When I draw a connection on a page, the line is always intermittent with breaks where the pencil lead and the bumpy page leave gaps. Even when I conceive of a connection and can imagine a line that appears continuous, my mind focuses on it intermittently in the quanta of neurological connection. There is no continuous connector; anyone who thinks about the same thing for a while will notice that intermittent character of connection.

And here we are! The presence of multiple syntactic formations in poetry (all cultures, for that matter, both because of the diverse needs of people and because all languages have similar capabilities even though they may be ordered and expressed differently) facilitates it as a social force operating like formations in the biosphere in pursuit of problem-solving for needs like energy, food, social cohesion, and cultural vitality.[12] Under the yoke of slavery and oppression, Black communities have developed multiple cultures in some way independent of their white counterparts. Although the influence of white male ideals in film and television has global reach, white suburban Americans and Afghan rebels high five, and African rhythms echo in music throughout the world.

I'm standing outside Citibank's Emerald City building talking with some of my workmates as they smoke cigarettes. One of the women says to me, "James, your only problem is that you weren't born Black." "Yes, I know. Jews were isolated from Europe for two thousand years. Given a bit more freedom and a patch of land, Israel reproduced the separation from its neighbors. Go figure."

Parataxis and hypotaxis! An environmental model includes parataxis to emphasize our non-linear connectivity and social conditions. Hypotaxis supports individual expression, the nuances and argument of the environmental model. Separately they emphasize political and social values of connection. Together they help create a fuller expression of self than either one alone, emphasizing individual, social, and biological networks.

Scalable Syntax: Poetry Model of the Biosphere

This chapter extends the discussion of the scalability of environmental networks by citing how the entire biosphere scales, including in language, from the smallest to largest with obvious discontinuities and arbitrary intervals. Some scaling is only local and specific like the sizes of Matryoshka dolls. Some scaling is large and general, independent of size, that is, isomorphic. The chapter shows how to think about what is similar, what scales, and what is different in a multi-language poetry ecosystem that progresses poetry from the writer to the reader and throughout the network in poetry creation, writing, translation, publication, and distribution.

* * *

Scalability in the biosphere can be universal. The components that build life have related structures at the cellular, organismic, group, ecosystem, and biosphere levels. Hyunju Kim et al. conclude:

> The application of network theory at a planetary scale can uncover properties existing across different levels of organization within the biosphere and can be predictive of major divisions within a given level (such as evolutionary domains).[1]

J. Sherry, *Selfie*, https://doi.org/10.1007/978-981-19-4870-1_12

Language structures also scale throughout the biosphere in:

- genetic sequencing (A, G, C, T),
- human and non-human communication,
- human thought processes in the way both logic and the subconscious are structured linguistically,
- social networks connected through language structures as discussed in the prior chapter,
- engineering through mathematical language and people's descriptive use cases, and
- global organic processes like nitrogen cycle signaling pathways.

The sequences of syntax can be understood to scale to event sequences, that is, what occurs to bodies—human and social bodies, bodies of water and land, bodies of thought. Events also scale in poetry when using writing, translating between languages, publishing, and distributing works of poetry. All these processes have linguistic properties and are also structured in part through language. In the recent study by Hyunju Kim et al., scaling has been shown as universal in the biosphere:

> Biology exhibits some of its most notable regularities moving up in levels of organization from individuals to ecosystems, and these regularities may only truly manifest at the level of ecosystems and, ultimately, the biosphere. For example, while individual organismal lineages fluctuate through time and space, the functional and metabolic composition of ecological communities is stable. To understand the general principles governing biology, we must understand how living systems organize across levels, not just within a given level.[2]

Poetry supports further understanding of ecological networks when it focuses on these scalable ideas like syntax that scales up to the general concept of sequence and then to any of the formats of ordering/organization. Scalability clarifies isomorphism and connectivity between individuals, groups, and surroundings as well as between disciplines of thought. Michael T. Hannan and John Freedman: "In any given environment (in equilibrium) there will only be one form of organization (isomorphism),"[3] similarity across scale. Related to Baudelaire's correspondences, isomorphism includes adaptation at multiple scales and inertia further nesting both change and the difficulties of change.

Poetry's syntax, translation, publication, and distribution all participate in similar step by step adaptive/inertial principles of correspondence, one kind of scaling. Words build up to phrases. Functioning tends toward assemblage and the process of construction. Ecological studies of the biosphere show how things are similar ≈ at many scales. Some are asymptotic ≈ (approaching), some are rangy ≥ , and some only relate in a few components of their identity. Carolyn Merchant:

> Theories about nature and theories about society have a history of interconnections. A view of nature can be seen as a projection of human perception of self and society onto the cosmos. Conversely, theories about nature have historically been interpreted as containing implications about the way individuals or social groups behave or ought to behave.[4]

Writers start by listening and reading. Children don't separate themselves from their parents, mostly listening, until about the age of five. When in the middle of reading him a Harry Potter book, my son said, "Dad, I can read it myself," the character of my own connections to him changed and I had to redefine the linkages between us. For him, it was a major victory, enhancing his own theory of mind and independence. One version of events might be described as Oedipal; another version might be developmental. Both operate as part of the same hierarchy of self but through different disciplines of thought. My son's assertion made it clear that he understood how I thought. By removing himself from an identity refracted through me, he understood more about himself. He was able to interpret the intention of many of his friends and texts. He read books and listened to speech as they connected into the way he used language both written and spoken. From earliest childhood, the syntax of his growth followed the books on the sequence of childhood development detailed by T. Berry Brazelton that my wife and I as devoted literate parents read.

Once such children separate from their parents and their childhood surroundings, they sit in rooms surrounded by books. The alignment of the books on shelves and the nested organization of texts online infiltrate individual thought, creating a selfie with books in the background, confounding the childhood hierarchy of thought and introspection.

Climate change intrudes, too, and the anxiety it provokes begins to distort perceptions of individuals, communities, and nations, raising questions about the evaluations and prerogatives that people have accepted for hundreds of years. Both children and adults focus increasingly on their changing interior condition and narratives about themselves, driven by global warming, even though looking outward would seem a more appropriate way to combat climate change. Social media also distort this interiorized focus: you can tell thought processes are changing because of the extreme emotional reactions to climate change, especially on social media.

Nevertheless, because many networks in the biosphere, as researched in a "global database of 28,146 annotated genomes and metagenomes and 8658 cataloged biochemical reactions"[5] composed by Hyunju Kim et al., are nearly scale-free even as discontinuities exist, looking at oneself provides some accurate information about world and word function, if not the specific actions required for climate remediation. Solutions to climate change, as the environmental model proposes, lie inside and outside ourselves through the networks that sustain life in the six industrial processes that intend to secure and expand access to food, clothing, shelter, and cultural interactions. Isomorphism makes inferable information available on many scales in spite of the distortions about the world that looking at any self-like thing by itself, alone, engender.

Within these restrictions, imaginative children insist on their own agency, no surprise, by apparently not scaling. Inventiveness imposes what it perceives as its own organization on received sequences of power, reordering bodies in spaces, testing which narratives scale within hierarchies created by their observation and parental discipline, and breaking into and breaking down borderlands around themselves. As young adults, feeling the shape of their social world, they resist, protest, and exclude to form new connections (still isomorphic and scaling) as further affirmation of self, group, and surroundings. I remember supporting my son's interest in reptiles with books, pets, and visits to zoos. When I gave him a more advanced book on reptile biology, he lost interest, feeling oppressed by the details he had difficulty grasping. His reaction made me aware of the difficulties of any sustained social disclosure like the multi-scale realignment of assumptions that might slow climate change. These difficulties point to the way poetry has to work together with other methods to actually be effective as an agent of change. But as people age, they increasingly accept the majority of what they are given and follow patterns, and poetry effectively creates language patterns.

Poets' identities and styles, however immediate and innovative, are built around books and poems as models and recently around computers as much as tranquilly remembered sequences. Some poets speak about writing word to word, some about writing poems, some write only poem to poem, some poets think through writing books, and some think about performance, instrumentally using language to shape events. Most consider all of the above at different times, but each of these represent priorities in a part of the poetry process that climate change might alter.

The order of development is similar at these different scales of reading, writing, and recalling for mass market and for narrowly targeted audiences. The weight of knowledge and the ecosystems it inhabits drive it into specific disciplinary and even individual solutions that often fail to travel across the connections that support them, locked into marginal forms. These formats are connected to each other both in theory and in practice, while they operate at different scales in the ecosystems of poetry, they can all be described as methods by which poets develop their ideas, style, and language.

Poetry then needs to be understood as a network of remembering, listening, reading, and inscribing where some or all of them occur for each instance. Isolating one or a pair from this syntactic sequence may be good for argument and easy to track, but following the network of linkages and nodes models poetry more accurately as a process. Without identifying connectors, poems fail to complete the environmental modeling process to become comprehensible.

This network of writing performs a model of writing prior to beginning to write, that is, everyone has an idea of what writing is before they sit down to write even if it's not well-formed. Sometimes that model is improvised but poetry is still written with a prior model. The network allows many different varieties of writing. This network develops different models similarly for each writer, that is, again, isomorphism, where you and I have both general similarities and specific differences. Evaluating a poem is rarely more than asserting yourself, but important commentators build models of analysis based on psychology, social theory, observation, and ecology and apply them to poetry.

As an example of environmental change, Shelley simplified: "Reason respects the differences, and imagination the similitudes of things."[6] In showing poetry's complexity as a poet/ecologist 200 years later, Shelley's binary changes so that reason supports both differences and similitudes. Imagination tends toward connecting things, but also encourages the poet to assert their independence, if for no other reason than to feel good about their writing.

Poetry also scales beyond these few categories. How many levels are significant? What constitutes a difference of signification applicable to changing how corporate leaders view their surroundings? In the environmental model, the levels of signification vary with conditions and requires collaboration with other disciplines of language. It may mean that some of the cherished identities of poetry may not persist and are not sustainable, although they can be recovered in archives. What will be significant will include our surroundings and the network that sustains ourselves and our species. Demonstrating this additional component through the many channels I'm trying to introduce in this text will be crucial if poetry is to help to slow global warming.

From the environmental perspective, poets translate together in several possible configurations and through several possible channels carrying more or less communication between the poet and the translator(s). Beyond poetry, translators between languages are not isolated from or even that dissimilar from human activities at other scales and so represent a good example of the environmental model. The linkages are complex but translation is ubiquitous from language to language, from perception to language, from thought to language, and from memory to language. Thinking like this encourages poets to look at their work more like ecology and less like an assertion of individual separation and distinction. Of course, this remark may antagonize poets adhering to other poetries.

Ecology's most contentious assertion in my experience has been that writing takes place, in part, through social groups. Sources and practices of writing and translation operate in minds, on pages, in language, and through social relations. Poets also write from a location that can have greater or lesser impact on what they write. I would call this map of poetic activity ecological.

Here's an extended example. In a new book of translation that I hope eventually to publish, *The Reciprocal Interpretation Project*, Sun Dong and I publish poets commenting on each other, adding yet another level to translation. In the book, six Chinese and six American poets interpret translations of each other's work. All together build an ecological network of poets, poetry, and writing about them published and distributed to the US and China.

Here is the matrix of who will interpret whom, the network model. For example, Rae Armantrout will interpret Duo Yu, Zhou Zan, and Hu Xian. Yu Jian will write about Rae Armantrout, Tan Lin, and Anne Waldman.

	Yu Jian	*Che Qianzi*	*Duo Yu*	*Zhai Yongming*	*Zhou Zan*	*Hu Xian*
Rae Armantrout	1		2	1	2	1,2
Ron Silliman	2	1,2	1,2			1
Tan Lin	1	1	2	1	2	2
Divya Victor		2	1	2	1	1,2
Anne Waldman	1,2		1	2	1,2	
Lyn Hejinian	2	1,2		1,2	1	

1. Means Chinese poet interprets US poet
2. means US poet interprets Chinese poet
For example, Zhai Yongming interprets all poems by Rae, Tan & Lyn

This matrix makes clear that all participants are required for the book to work. No one is marginalized. There are no extraneous or politically expendable writers. All must participate. This model then is more extreme than in the biosphere where often redundant systems make components expendable. In the human body, there is only one liver but both the spleen and bone marrow make red blood cells. We have two eyes and two hands. This structure not developed by logic but is functional and developed by trial and error. Another trial produced eight-armed organisms.

Chinese and English poetry have different bases due to the different ways meaning is transmitted through Chinese characters and English words. Characters and words are not always the same level of discourse. They also emerge from different cultural environments and different geographies, but translation enables us to speak to each other, albeit imperfectly, within ranges of comprehension. This approximation through an environmental model replaces the absolutes, evaluations, and essences of humanist poetry in both countries. Approximation operates at a larger scale than the poetic ideas of suffering, sensitivity, and skill as creative forces. Suffering sometimes occurs because of unrealistic expectations of perfection. Sensitivity varies from moment to moment and between locations even for the same writer. And skill, if it is linguistic skill, selects appropriate poetic strategies, forms, syntax, diction, and tone that match or contrast with themes.

Is there any single strategy for understanding each other or are there many valid approaches? I'm suggesting there is not one but many. One of my favorite books of translation is called *19 Ways of Looking at Wang-Wei.*[7] In this book, the American critic Elliot Weinberger and the Mexican poet and diplomat Octavio Paz built an ecosystem of translation with different species of translating and different translations of the same Wang Wei poem over several centuries. Sun Dong and my reciprocal translation process extends the multi-perspectival approach of *19 Ways* into a wider ecosystem of forms and connections but a narrower one of time in that we're only working with living poets translating and analyzing each other. Highlighting these similarities and differences points to an ecological, that is, connective view of translation.

At another scale, in this ecosystem of translation, a network of synapses allows us to read. This network performs multiple translations as well as multiple types of translation: literal, neurological, chemical, conceptual, evocative, and transporting. The network expands into channels for distribution around the body. At different scales, the same processes occur. I want to point out again, this is a functional description of ecology that poetry and distribution mimic—biomimicry.

These scalar networks of poetry and translation are built using resources of many types. Intellectual resources are required to design the network, financial assets are required to build out physical networks of distribution, and poets communicate to each other over the internet. All of these network nodes and channels require human labor, machine labor, and resources from our surroundings to create and maintain, what E. Summerson Carr and Michael Lempert call,

> the way different potentially scalable qualities or dimensions can be made to reinforce each other, almost like a kind of scaffolding on which people rely but take for granted.[8]

Empires are notoriously poor contributors to translation because of their assumption of transnational control and central importance that tends to ignore peripheries. Networks of US translation sadly often collapse because of the lack of money and other resources applied to translation in a system dependent on transnationalism, capital, and expansion each at its own self-reinforcing scale. Predictably, the translators respond in kind as part of their culture. Sun Dong and I in our reciprocal translations think we can learn something about each other by looking at the same scaling of poetic assumptions from the point of view of the other culture.

In the environmental model, we need all of these resources, not just some of them. Intellectual resources alone are insufficient for translation as other subnets are required for distribution. In the US, many people are using their intellectual resources and applying labor to it, creating lovely translations that are largely unread. Without support by educational institutions, books of translation languish.

Because limited resources are applied to translation, different methods of translation remain isolated and contentious. The lack of resources causes conflict and reduces free flow across the network. This is both perceived and real. The competition for limited resources—the same kind of problem that we see at the geopolitical level, at all scales of human endeavor—occurs at the poetry translation level, where poets compete for access to the few resources available. Poetry writing, initial poetry writing, does not really require such significant resources, just me at my desk with a pen and paper. But translation more than doubles that effort and requires far more resources because two people or multiple people work together and transmission resources like mail and internet are engaged in the translation process.

Distribution networks also require money and labor, but because there's a perceived return on investment, money and labor are applied to distribution. China has strong regional networks and the US has national networks of book and poetry distribution. Here are some statistics:

The US Postal Service shows us how to distribute 1000 copies of a book of poetry in translation, which is about the number of books that we distributed in the US of the *Reciprocal Translation Project*. There are 25,000 post offices, 215,000 mail trucks, and other relevant statistics. But there is only one cost-effective way to send books and that is called Media Mail. Media mail in 2020 costs $2.66 for one pound or less. For every additional pound, the media mail cost increases only 51 cents. So, the more books you send at once, the cheaper it becomes on a per book basis. This is basic economics.

In China, similar but different institutions operate. Postal offices and branches, many more than the US: 82,000, but about the same per capita. Mail processing centers two hundred and thirty-six. More humans labor in the system of distribution. But, in China, access to the ecological network of publishing (network gate) is article 11 of the regulations of the administration of publishing that states anyone establishing a publishing unit must have registered capital of 300,000 RMB and or about $35,000. To become a publisher, you need that kind of investment capital. In the US, we don't have a strong financial model of publishing for poetry or translation, so this creates a major difference between the poetry model in the US and China.

With all of this infrastructure, you'd think dissemination of translation would be easy. All we need to do is find prospective readers, contact prospective readers by snail mail, phone, chat, or email, and send books by post. The best approach, however, is what we call "sneaker net," transferring by hand at readings or over coffee. The Post Office is less effective. We could send PDF files to readers through emails, but most in my experience want a printed book of poetry and once my publishing company spends the money to publish, it wants enough returns to publish the next book. So, the publishing institution has resource needs like individuals.

Scaling in these networks operates similarly. As a publisher and editor, I realize how the ideals of poetry life operate through what is practically achievable. What is considered "practical" primarily illuminates what appears able to be engineered institutionally across the networks available in the US and China. But as an alternative, and assuming scalable networks, the political right in America does not need to be told after forty years of institution building that what is practical also must be scaled up over time.

Finding how culture can influence climate change at different scales needs that kind of stepwise learning about what can be accomplished in a given ecosystem and pattern of intellectual climate. So, the ambition of these books of translation appears not through sales but through imagining how poetry processes can slowly influence thinking about environmental issues. These books of translation are an environmental project and can best be understood through looking at them ecologically instead of evaluating each component—original, translation, collaboration—separately, although that's also easy enough to include in the process of reading.[9]

But even with these grandiose objectives, translation sales are minimal. At the current levels of volume of sales versus cost, most books of translation lose money. This leads back to a negative economic value of poetry. As I have often said:

You can buy a piece of paper for a penny
But if you put a poem on it, you can't give it away!

This frequently cited poetry humor hides a difficult value structure for poetry and poetics in the US. My environmental work proposes a value structure for poetry and translation quite different from the current model, yet I persist in assuming that some of these ideas will synchronize with other projects at other scales to help change culture toward an environmentally friendly model.

The politics of distribution across borders represents a further, significant barrier to the dissemination of poetry. Customs and border control in each country have a significant impact and cost. Last year, a friend shipped a box of books from Serbia to the US by first-class mail. The box never arrived because she failed to pay the extra fee required to get a customs declaration automatically sent to me so that I would know it had arrived in the US. The books are lost in some warehouse in the bowels of New York City airports. The environmental issues are capacious and capricious.

Customs and border control monitor and act as a gate for distribution of books between nations. Postal services, intra-institutional delivery, and private package delivery act as another part of the social landscape. In the US, privatized delivery has dominated the national postal service even though the postal service is far cheaper. DHL, United Parcel Service, and Federal Express: their franchises have been protected by our government with borders, and crossing borders costs money to which they add a charge. The same one-pound book, sent from the US to Brazil, costs $120 via DHL, versus one pound via postal service to Brazil costs $32 or about 12 times more than the same transportation within the country.

Without media mail as an option, first-class mail of a one-pound book from the US to China costs $16.75, approximately equal to the retail cost of the book. Private freight service from New York City's Chinatown costs eight dollars a pound to central China. Considering the distance to China, freight delivery is the most efficient way to transmit translation between China and the US supporting the value of globalization rather than fortifying borders. This is a step in the right direction, but for these books underground networks operate more cheaply. Shipping from the US to China costs much more than vice versa. Volume and cost of living may be relevant to the difference, but also US and Chinese governments' priorities are different.

Finally, and briefly, consider the availability of paper and ink. The forests of the world are being depleted for books and magazines, to make way for grazing land for cattle cum hamburgers in Brazil, and for people who want to move "closer to nature," one of the prime causes of financial losses from wild fires. Yet, trees are a renewable resource that can be managed to support sustainable paper, meat, and land availability. It's not binary, but does require significant attention to detail.

With all this infrastructure and other factors such as wide access to texts and images on the internet, the average American reads four books per year. 30–40 book distributors in the US enable this through mail and freight delivery. All major book publishers are also distributors for their books and often other publishers. The Chinese public reads about 5.5 books per year distributed through regional networks. In both countries, the underground press flourishes along with individual printing and dissemination, private readings, and talk groups. Universities become safe places for non-commercial poetry and publication to flourish. The ivory tower protects poetry as well as operates through its own principles of evaluative hierarchy. Increasingly, considering the rising per-volume cost, students don't read books of poetry, but just read individual poems in anthologies or on the internet. There's a risk to the university model, however, when fewer students pay tuition or no longer seek cultural education versus scientific and business training. Academic gates such as value judgments also inhibit poetry in universities.

In the past, within the Anglophone network, editors acted as the gate keepers of poetry by determining what books readers would have available. Now, with much lower cost publishing technology, distributors have become the gatekeepers of poetry. If you do not sell or distribute enough books, the distributors cannot make money and will not distribute your books. Without distribution resources, there is reduced availability of poetry for readers, except by hand, mail, and email. Thus informal, underground networks operate effectively in the US for much poetry. There is direct marketing via Amazon and various internet (self-)publishing services seeking to replace these informal networks with individual to corporate network to individual connections.

Selected individual poets move into commercial literary networks, but their numbers are few and while distribution may improve, publicity budgets are small. Academic affiliation adds capital to publication and distribution as well. US and Chinese academics writing poetry can travel for their jobs providing additional channels for distributing their work. The institutional nodes support improved financing of flows of information in both directions.

From the environmental point of view, as we travel up and down the scale of networks, we can view distribution's role in the syntax of completed works of poetry and translation. In my search for scalable ideas, I suggest that syntax: texts = distribution: books. Without the proportion, you can just say syntax and distribution are similarly related to their resources. The isomorphic details of their similarity are not, in my view, any less significant than the way in which syntax facilitates the meaning of English language poetry and translation. There are far more critics of poetry texts than distribution methods so the ideas around texts are more thoroughly investigated than distribution methods.[10] These few distribution methods are more fully vetted through practice and success as determined by principles of fitness in marketplace, finance, and technologies of distribution.

These networks of translation and dissemination build multiple translation models as thought moves through gates and routings, a kind of social thought outside the mind where groups act as a repository of knowledge. In addition to thinking outside the mind,[11] network connections between individuals, groups, institutions, and their effects on surroundings are carriers of thought. These translation links operate at the level of neuronal thought, speech, editing publications, translating between languages, and distribution of publications. This thinking with translation creates diverse networks. Translation of thoughts, speaking, editing of texts, distributing/publishing/reading books, and searching responses by individual readers, professional critics, and a developing canon build a specific model that looks much like the ecology of ecosystems.

Distribution networks build and connect multiple translations in the same way ecological networks build and connect multiple species through taxonomic categories—a kind of hierarchy. In addition to thinking outside the mind, connectors actually carry thought throughout the body. Neurons facilitate speech and writing that I can communicate back to you. They enable editing and distribution of poetry, translation between languages and distribution of publications throughout the world. The translation of thoughts, speaking, editing of texts, publishing/distribution/reading books, and searching responses by individual readers, professional critics, and developing canons are another part of the ecological model of thought. In this ecological processing, translations make connections. Connectors between people, groups, and surroundings also translate how people think about themselves. Thinking outside ourselves, between us, our distribution conversation is a kind of thinking about how we regard/consider/handle/treat each other: ecological construction of translation.

We compare phenomena and organize our world through the network. Whether the network hierarchy operates through the evaluative criticism of good, better, best, or through the inclusive hierarchy that looks like the Linnean taxonomy of species, distribution, writing, and thinking, all operate similarly through linked networks. Operating at many scales, multiple translations produce a condition, that psychoanalyst Yolanda Gampel called "interminable uncanniness."[12]

Post-Traumatic Stress Disorder produces interminable uncanniness.

Climate change produces interminable uncanniness.

Poetry at its best produces interminable uncanniness.

Translation in its oddities produces interminable uncanniness in the sense that the new version is never quite what we expect, even less so when processed through a complex network of distribution rather than only between two individuals.

The Anthropocene: Ecosystems & Time Frames

This chapter restates the definitions of some environmental terms as multilevel and conditional. Many of the terms from ecosystem to Anthropocene have several definitions that change with each writer, their discipline, and situation. Definitions tend to be vague and need to be developed as composites with other concepts for accuracy as pointed out with the concept of freedom above. So far, these differences have mostly promoted disagreement instead of recognition of the value of difference under different conditions. Descending from the great ages of earth to recent definitions of concepts like ecosystem, the chapter shows how culture adapts to similar approaches to problems regardless of scale, that is, how culture adapts to climate.

* * *

The concept of the Anthropocene, proposed as part of a fixed historical sequence of geological ages, links the biosphere to how the individual thinks about their species in time and helps follow the trajectory of human effects on the biosphere. Disagreements persist regarding when, exactly, the Anthropocene began some saying as recently as 1945, some 1780 (near the beginning of the industrial revolution), and William Ruddiman[1] pointing to the beginning of slash and burn agriculture 8–10,000 years ago. Simon L. Lewis and Mark A. Maslin point to a "dip" in atmospheric CO_2 dated around 1610 as a signal of reforestation of the Americas in the wake of colonial genocide and reduction in American native settlements.[2]

J. Sherry, *Selfie*, https://doi.org/10.1007/978-981-19-4870-1_13

Focusing on these multiple views of the causes of climate variation, climate encourages me to think about climate as naturally fragile as the climate of human thought. These conflicts about the human effects on climate improve the visibility of concepts like the Anthropocene because the mainstream media would rather cover a contest than interpret. The fight reinforces the organismic binary, its ego support, and the misunderstanding that human activity is destroying the planet when in fact it's primarily destroying human civilization and species relying on similar ranges of heat and water.

Integrated into and performed as an ecosystem in itself, environmental culture, including climate science, poetry about ecology, and environmental protest movements, finds use value and ethical value in the concept of the Anthropocene that change current accepted notions of the individual in the developed world. The idea of the Anthropocene conditions a focus of yourself toward our species. The phrase implies that something needs to be done at the species level and that that something will be different with respect to the biosphere than what is being done now. Poets seeking connections to the past may find the concept of the Anthropocene both useful and maddening because it attempts to treat all humans similarly. At the risk of repeating myself, humanity as a whole has not caused the climate change that is increasing. Six industrial polluters are causing climate disruption. Trying to point to all humanity misses the point of the cause, although the sensitivity of the biosphere is highlighted by Ruddiman's analysis of the early impact of agriculture.

Not all cultures and geographies are implicated in the same way in either the causes or remediation of climate change. Therefore, it doesn't make sense to treat humanity's obligations as identical for all global cultures, and the value of the Anthropocene might be primarily useful for developed nations and then not for all their citizens. If it's helpful at all, the Anthropocene chides developed national institutions and leadership of states and corporations. Eduardo Viveiros de Castro and Deborah Danowski:

> In the context of climate catastrophe that defines the Anthropocene, the line separating victims and culprits is historically clear from a collective or societal point of view, but much harder to trace from the point of view of individual action. This is because, today, many of us (humans and the nonhumans that have been enslaved or colonized) are victims and culprits all at once, in each action we engage in, at the push of every button, with

> every portion of food or animal feed we swallow—even if it is as obvious as it is essential that we do not confuse McDonald's itself and the teenager conditioned into consuming junk food, or Monsanto and the small farmer obliged to spray his genetically modified corn with glyphosate, let alone the pharmaceutical industry and the cattle force-fed with antibiotics and hormones.[3]

Relying on the concept of the Anthropocene to change culture ignores this fact: most people don't see themselves implicated in the Anthropocene, but do blame leadership, corporations, and perceived leadership. They blame their neighbor's leaf blower but not their own SUV. Moore:

> the growth of our understanding of how humans are made by the rest of nature, and of how nature is made by humanity has stalled. Nowhere is this clearer than in the popularity and influence of the dominant Anthropocene argument… Humans are placed in one category, Nature in another, and the feedbacks between them identified…. But such perspectives pose a deeper question they cannot answer: *How do humans co-produce patterns and relations of power and production within nature?* The question cannot be answered in a dualist frame.[4]

As people in the developed world become more aware of the academic concept of the Anthropocene and its implications, distinguishing between humanism and environmentalism becomes more vital to changing minds about how to cool global warming than adding to a geological timeline. The humanist model of the self, discussed in the Introduction as embedded in society and in turn in the biosphere, is implicated in the current agents (people, assemblages, and products), conditions, and situations causing climate change, including exclusionary separation of disciplines of thought, capitalism, and exclusionary race and gender politics. Models drive more deeply into people's consciousness than naming because they operate at many levels while concepts like the Anthropocene seem easy to ignore.

People hope for something new that improves their well-being, but the Anthropocene implies an end point for human domination of the planet. Environmentalism offers renewal, superimposed, rather than dragging our tattered humanist baggage into the next era with endless justifications of increasing social complexity.

Finally, the fact that the concept of species doesn't scale up to the biosphere further reduces the value of the Anthropocene in changing industrial behavior. But species scaling is more nuanced than some commentators suggest. For the concept of the Anthropocene to be effective, we must, as Dipesh Chakrabarty says, "'think of human agency over multiple and incommensurable scales at once.'"[5] Derek Woods himself then says: "At its most general, scale variance means that the observation and the operation of systems are subject to different constraints at different scales due to real discontinuities."[6] I can observe scalar similarity, but systems may operate quite differently at different scales. A species for example operates to reproduce itself by using resources, and that scales down to individuals, but the operation of the ecosystem where species live might constrain those resources. And at the biosphere level, there is no species as a comparative, only the singular layer of life over time.

Without a scalable species, the concept of the Anthropocene doesn't address the problem as corporate leaders and specialists change the assemblage. The species isn't remediating the climate, the assemblage composed of groups and technology does that. The environmental model does show similarity at different scales. Woods: "Scale critique shows that the subject of the Anthropocene is nonhuman."[7] Yet scalability outside of the species, scalability with discontinuity remains critical to understanding how to present the problem of climate remediation.

The term Anthropocene also categorizes time in a singular, geological manner, one of the varieties of time listed in the first chapter: atomic time, sensory time, observer time, biographical time, historical time, biological time, geological time, terrestrial time, and cosmic time. The environmental model on the other hand benefits discourse around change by speaking about time as a proportion through these multiple dimensions.

Even time for individual observers operates in different overlapping and subdivided tenses. Past, present, and future each have multiple subordinate tenses. Present tense includes present simple, continuous, perfect, and perfect continuous, but the present hardly exists. We transition through it from the past to the future. Our consciousness is always looking at the present, constructing it if for no other reason than our senses connect in near real time. Our lives hang by a thread that is most visible in the present. Fear of the Anthropocene, fear of power, is asserted in present tense even though it points to future events goaded by prior acts. We attend dominantly to that variable density present which makes it difficult to see ourselves accurately since the present is simultaneously transient, continuous, and fragmented by attention. Here's a view of the individual, social, and environmental time (Fig. 1).

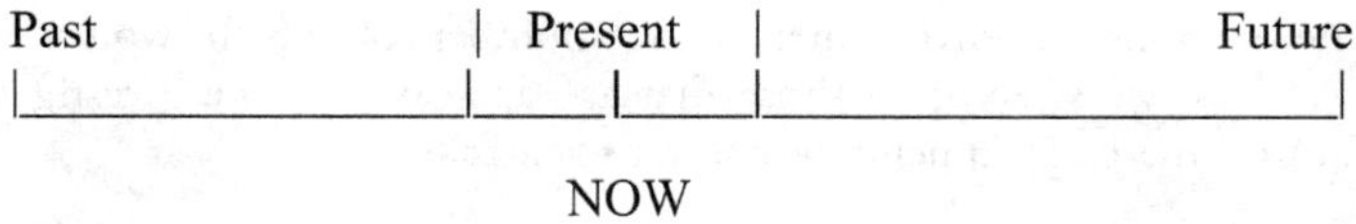

Fig. 1 Representation of NOW in 3-D time scale

The ways of acting and thinking proposed in the environmental model as propelling progress are mostly focused on nation states financialized by late capitalism, driven by corporations with government backing in the US, Europe, and more recently China and India, who are primarily responsible for climate change to date. These are the corporation-dominated state cultures that need to change. Indonesia with massive slash and burn pollution, Nigeria with oil production, and Brazil with deforestation are significantly implicated.[8] All state cultures, based on the complex inter-intra-actions between the need to care for their populations and the demands of their oligarchies, have the propensity to go through this change and all must reroute their traffic and inform their leaders about change.

* * *

The disciplines of thought have been radically subdivided in the past century on the mandolin of discourse. Critical theories centered on the inner life of the individual and on social life will integrate the third term—environment or ecosystem, habitat, surroundings, setting, domain, location, place, position, and duration—into their perspectives. World ecology that rejects the nature/society dualism will be enhanced by including the details of ecological networks as they operate for habitat, society, and individuals. Humans as the centerpiece of the biosphere move in a Copernican fashion but within taxonomy to another set of locations. The new configurations will be many and likely subject to their ability to integrate with the biosphere as it too radically changes.

Like the multiple view of selves there are multiple ways that environmentalists look at their terminology. Even among professional ecologists, terms like environment vary in meaning. For von Uexküll, each animal has its own environment (habitat and also related to the model built by Michael J. Reddy) and its own view of the world where as von Uexküll points out:

> Of the millions of environments, the number of which would only confound us, we seize upon those that are dedicated to the investigation of Nature—the environments of natural scientists.[9]

He goes on to speak of "the environment of astronomer[s]" and even the environments of individuals where John M. Meyer and P. Wesley Schultz have different environments much like the popular view of multiple universes (multiverses). Von Uexküll shows how the "worlding framework" *umwelt* looks different to different professions:

> In the behaviorist's environment of Nature, the body produces the mind, but, in the psychologist's world, the mind produces the body.[10]

Environmentalism suggests the connection model is at least bidirectional in every case even if one direction is seldom active.

For climate activists, on the other hand, environment equals the world. Roy Morgan, a mainstream Australian market research firm, produces surveys such as one entitled "Australians see the Environment as the World's biggest problem..."[11] Environment there is equated with the biosphere and biolayer as a whole. The word environment because of its multiple meanings acts like a metaphor rather than a literal term as if any language could be literal.

Beyond the ecosystems we transit, in his article "Call the Pulsing Home," Jonathan Skinner speaks of:

> language, as an ecopoetics, entails a systems-based rather than a values-based understanding of environment—one that does not posit 'the environment' as background to human activity. Rather, distinctions between self and environment are internal to human and, indeed to all, forms of life. What we call 'the environment' emerges constructively, to the extent that we sound and are sounded by the pulsing of these myriad life forms.[12]

The systems view of environment that Skinner refers to, Nicklas Luhmann's *Ecological Communication*,[13] links to the autonomous view of nature. In an autonomous view, a person's environment is accessible only through the individual's history and perception which to Luhmann means programming of the system, a perspective not dissimilar to work on metaphor by Lakoff that appears in the next chapter. While some systems theories and capital-based solutions address the causal problems in ecosystems with engineering—carbon capture, reflective surfaces, large scale solar, wind, and geothermal energy plants, et al.—other systems theorists understand the problem as one of communication and building consensus in order to propel fixes into the larger population as a global culture that also retains control by capital during the consolidation stage of human evolution that appears to be part of curbing our population bubble. Still others intend systems theory for an "ecological communication" to break down and reconfigure how people view their ecosystems.

Vernadsky attempts to provide an early but complete view of the biosphere as a "living organism" that

> should now be studied empirically, as a particular body that cannot be entirely reduced to know physico-chemical systems. Whether it can be so reduced in the future is not yet clear.[14]

Life establishes "a dynamic equilibrium, not unlike the evaporation of water from its surface…"[15] This kind of autopoiesis was reproduced more recently and independently by James Lovelock and Lynn Margulis in the Gaia Hypothesis.

The most apparent rejection of biological and organic determinism derives from their past use in eugenics and racism. Are biological models then forever lost to our cultural research because of past misuse? I would hope not, since we are made of organic compounds, eat them, and breathe because of their chemistry. In riposte to the redoubtable Barad, including nuanced biological views of environment and culture has more benefit than risk. The body is a sine qua non, but its influences include social structure, culture, habitat, and assumptions derived from the manifold. We cannot anchor patterns of gender, skill, and value in biology, but it does have an influence on society, habitat, and the ecology of relationships.

The social ills generated by biocentric arguments in the past can be addressed by legal and cultural adaptation. As mentioned above, any position can be misused but we cannot ignore our bodies. Histories centered in the body and on documents make the same essentialist mistake. Narrowly centered histories are useful because they can focus in detail, but the categories create other problems. The body remains important in personal and social life. My body's derivation from and influence on the biosphere with its countless bodies and forms of bodies throughout the layer where life operates spreads both into me and into my surroundings.

As Deleuze and Guattari point out about philosophy: "the concept is not object but territory."[16]

The body is the territory we inhabit unless the reader proposes some version of metempsychosis. Although the self extends beyond the body for reasons of ecological repair, the self certainly has a lot to do with the body. We dispense with the body at our own peril, falling into the maelstrom of the disembodied technologies and spiritualism, sites where power remains forever in the hands of others. I continue to work on the specifics of exactly how to perform Gaia with geological, social, biological, psychological, and linguistic components. The different ways that metaphor works in the next chapter are an example.

There are those who think of animal as distinct from human, those who see human as animal, those who see them in common evolution, and those who see animal as living matter that requires protection and in some cases preservation in its current form. I don't think it's necessary to find common ground in a single global cultural view of both the climate problem and ourselves. Rather, for activism regarding global warming, multiple points of view and solutions, based in part on location/habitat, must be accommodated by environmentalism depending on the situation. The center may be framed as language, psychology, biology, society, and habitat. Although conflict persists without complete agreement as contradiction presents in the difference between individual perception and scientific observation, general consent seems to be a useful way to acknowledge the environmental multiple. General consent allows differences to win grudging approval of solutions within ecosystems managed by human governance that allows external influences to inform choice.

Networks of Metaphor

This chapter shows how metaphor works as a method for scaling meaning across networks. Metaphor exemplifies how poetry influences the climate of thought, connectors of language, and predispositions of individuals and groups. Metaphor is proposed as a figure of speech, a mental link, an agent of change, and as a demonstration of the similarity operating in the physical world because it reflects the inter-intra-actions of the components of the biosphere. Living things, their integration with the mineral and sub-atomic worlds, and the networks that connect them are all material as well as conceptual. Materiality of thought and language links world components. Identifying singularities appears as a satisfying if blinkered view once the reader accepts the material connections through which metaphor operates.

* * *

Few writers agree on what a metaphor is, but they know one when they read it. Some like David Harvey and then Donald Schön see metaphor as essential:

> …no matter how ruthless, pristine and rigorously 'objective' our method of enquiry may be, the framework of interpretation is given in the metaphor rather than the evidence.[1]

J. Sherry, *Selfie*, https://doi.org/10.1007/978-981-19-4870-1_14

> ...the framing of problems often depends upon metaphors underlying the stories which generate problem setting and set the direction of problem solving.[2]

On the other hands, the metaphor of networks has been widely criticized as vacuous by the likes of Mark Erickson and then W. Baldamus:

> The paper argues that the use of metaphors is probably inevitable, and can enhance as well as diminish our understanding of social experience. However, the network metaphor often creates artificial objects and makes us think we have been precise when we have been vague.[3]

> The paper argues that the use of metaphors is probably inevitable, and can enhance as well as diminish our understanding of social experience. However, the network metaphor often creates artificial objects and makes us think we have been precise when we have been vague.[4]

Taking Baldamus seriously, his critique of networks does make sense if social thought is isolated from the discourses around the other views of the planet. After all, why use a metaphor when the vocabulary around social thought in isolation from the other views of the biosphere is so well framed. But for the environmental necessities of climate change, I suggest that shared nomenclature is not incidental, but essential to link biology and psychology of individuals, social thought, and structures of planetary biology if we are to come up with a common model from which to diverge. Shared models help show differences and similarities between disciplines of thought, political frames and parties, and biological entities and systems.

Lots of room for disagreement and the multiple realities of climate. Since Reddy and others write vociferously against the accuracy and specificity of network metaphors, I have to ask whether the network model of the environment is a metaphor, physical, or both? The networks of the tree roots, nerves, blood vessels, and the internet are physical networks. And in this list, all are conduits. The networks of the nitrogen and water cycle are physical, but don't always travel fixed lines of distribution. Rather they are broadcast through the atmosphere rather than through a conduit. They soak through the soil and then follow well-defined circulatory vessels through of plants. Networks do not have to be conduits to transmit, but do embody connectivity.

Are the conversations among politicians about criminal behavior in the news media, followed by people hearing, reading, and talking about it, and then court documents physical? Do the intense feelings of love traveling between the eyes, loins, and pages of the poets or prayers to gods among believers while they look at the sky metaphoric or physical networks? Is the single complex ecology of humanity and nature that we label the biosphere a metaphor or physical? Is the radiation from the sun to the earth and reflected back to space physical? Are inter-intra-actions framed by the environmental model operating together through their networks that suggest methods for cooperative climate remediation physical or metaphor? In any case and to me, work/inputs/energy/labor is required for change of both metaphoric and physical views of ecosystems.

Reddy's concern is what he calls "the bizarre assertion that words have 'insides' and 'outsides'."[5] Wittgenstein makes the same kind of cognitive statement about insides and outsides. Looked at one way, words are units. But words do have insides and outsides, spaces, ways of shaping letters, and sizing them for clarity or emphasis. In affecting global warming, the physicality of words will be important. Understanding a word often calls for subdividing it into syllables that clarify its etymology. Words have prefixes and suffixes that change the word's meaning. The componentized insides with connectors, such as serifs and roots, as well as the whole word, move from the eye and ear to the cerebral levels of the brain both as wholes and as parts to be reformed for comparison and other cognitive processes. Linking a letter or syllable sound in one word to another phoneme or syllable in another word in a poem for a sensory effect in reading certainly clarifies how words are not monads.

Calling the assertion that words have insides "bizarre" hardly dissuades me from disagreeing. Rather it attracts my attention that Reddy feels so strongly that he needs to obliterate the thought; it must be worth investigating. Reddy's limiting the network model to a series of plumbing pipes, "conduits," is as partial as the web metaphor sidelined above. Erickson's calling the network metaphor vague is not the case here, because *Selfie*'s detail fills in the metaphoric network and examples of its physicality. I would assert that few ideas can be understood by themselves and need to be understood in terms of other ideas and other conceptual domains.

Words are physical and constructed as well as conceptual units. I can readily describe the connections of a word or phrase, spoken or written, to thought processes in the brain and back again to the next word or phrase, but the process is quite deeply and professionally examined through the work of Graziano and Joshua Tenenbaum later in in this chapter and connectivity is described further in the chapter "Identity's Constructive and Connective Ecosystem." What impresses me is the transparent way that language has allowed us to briefly describe how we understand words and their processes with a metaphor, a shorthand that is actually fairly accurate although usually within one discipline. Metaphors about how language works can be useful without being thorough.

Metaphor is one method that poetry uses to connect somewhat disparate things by a common attribute or two. The metaphor is also a connection. "All the world's a stage" reconnects the audience to the play after they leave the theater as well as connecting to the text. You can lose track of the connection, but it remains at the very least somewhere between memory and frontal processing. The metaphor increases awareness of connectivity. Can you write a metaphor that doesn't have such a reciprocal connection? Probably not because connection is inherent (natural) to metaphor.

Metaphor operates throughout language as Andrew Goatly points out from economics jargon like balance of trade to physics like free radicals (which also sounds like a political rally cry) to environmental concepts like greenhouse effect.[6] Metaphor inspires a huge literature describing and positioning it from Aristotle to Walter Benjamin to contemporary historical materialists. Metaphoric connections resemble social affiliation, family ties, and ecological relationships. Is the connection between society and poetry physical? In many ways, yes, but the environmental model addresses only a few of the interoperating characteristics of metaphor such as it is:

- Endlessly discussed physicality, that is, metaphor is part of material discourse
- traceable path from the biosphere to the mind of the writer to the page/voice then to the eye/ear of the reader to the effects on the reader's mind, a syntax and sequence of events
- methods transfer multiple meanings through the ecology of poetry to individual responses like surprise and tedium, social connection and environmental networks.

In the first characteristic, physicality, language operates the biosphere. Language shapes the genetic code (A,C,G,T) for all life and as, Jacques

Lacan famously pointed out, in the way our minds are constructed: "…the unconscious is the Other's discourse." And: "The unconscious is structured like a language." ("*l'inconscient est structuré comme un langage*")[7] We speak to each other and organize our societies through language. Humans are good users of this characteristic of the construction of the biosphere. Humans invent *and* discover language. We are good inventors of language, but we also discovered language in ourselves and in our non-human surroundings. (In this way, the pathetic fallacy is instantiated in the environmental model as a material function.)

The binary tendency of human minds may encourage the reader to evaluate whether invention or discovery is more important. Certainly, the Romantics focused on invention. But the process of writing is equally about discovering what works in the poem as one writes. In the biosphere, invention and discovery are both core functions as writers progress, and comparing value may be less important than accurately positioning each as operators in the network.

Syntactically, the environmental model proposes that the processes that generate images in the brain interact with the page through the eye and hand and then on to the reader. These processes resemble (approximate) other creative processes operating throughout the biosphere. Similarities between individuals and larger scales of life appear critical to generating most images that I can think of. These processes follow the network paths of the manifold of ecosystems built into evolution and thinking as adaptive agents. This similarity of form in process and things aligns poetry, speech, and thinking with the methods used in the soup of the planet to create life as directly and persistently as any theme separating the virtues of "nature" from human deficiencies.

Language is constructed similarly, too. The planet operates through interlinked subnets that carry phenomena and language. The subnets are phenomenal themselves. Language is collective, connected/connecting, and significantly scalable as above from genetics, to other biological constructions, to speech, to social constructions, and to the ecology of habitats. Like identities written and spoken languages operate at multiple scales. Metaphors make sense because language has its own ecology, speaking about one thing to connect to and imply others as much as connecting linearly through syntactic ordering. Language is not a human invention, although people have constructed a useful, complex version through our brain/voice apparatus, but how the world is constructed with humans participating in it biologically, perceptually, by hand and by alphabet and character.

This isomorphism of language and biology should not shock the reader. It should come as no surprise that language both communicates and performs within the person as well as between the self, their social, and their surrounding worlds. Language is constructed like the biosphere, that is, built up into grammars through different ways of using it. Metaphor functions using this ecology of discourse at many scales. In fact, when looking at the world from the perspective of language, language structures operate the biosphere. We have language structures in our genes, in the way our brains are constructed, in how we speak, how we organize our societies, and how our surroundings are organized both as we perceive them and as they in turn construct people as organisms. As a result, people are good users of this characteristic of the universe that operates in and for us.

The Five Graces Group suggests, language as metaphor adapts to change and also operates as an adaptive method.[8] Identifying the world as linguistic is not evidence of recursion, i.e., that the self is all that can be known to exist, but a fact of society and habitat built through connecting channels, of society as biological as well as political, of mind mostly organized like other entities in the biosphere, all adaptive. (And, of course, there are unique features at every level.)

As a figure of speech and in other ways, metaphor links one to another more through their similarities than their differences. In this way, metaphor follows Shelley's logic. The attractiveness and precision of any comparison may be generated by links to the organismic binary, by the need for familial connection, and by the physical connections that bind, perform, and communicate throughout connected ecosystems. There are two aspects to this configuration:

1. That which is similar is similar in a way that the reader's mind grasps quickly with resonances trailing through one or more levels in the brain. Looked at another way, metaphor can establish logical, anecdotal, and spontaneous connections.
2. The connections between Lakoff's concepts of "source" and "target" (or other binary) words follow maps or other grids in one or more of at least language, brain, body, society, and location.[9] The connections established in poetry for example operate though

a text/brain entity (Barad) even when the similarity surprises by its unfamiliarity like seeing someone we know in a new context.

The environmental model proposes that metaphor works through hierarchical mental processes similar to how social constructions and ecosystems operate. Starting with a simple two-level hierarchy of a social construct:

Level 0: construction site contains
Level 1: brick layer, carpenter, building materials

And of an ecosystem:

Level 0: forest contains
Level 1: plants, animals, fungi

Now consider three kinds of motion in this two-level hierarchy:

1. down to more specificity as the construction site contains a brick layer, a carpenter, and building materials while the forest contains plants, animals, and fungi.
2. horizontally within a level as a brick layer and the carpenter are both equally part of the construction site even though any one worker will consider themselves more important. Plants, animals, and fungi are on the same level of the hierarchy with respect to the forest even though they perform some different functions
3. upward to a higher level of abstraction. While it may be helpful for workers to know some facts about the whole project, it's often difficult to the brick layer to know his job as well as the schedule, purpose, and plans of the contractor and developer. Moving upward in the hierarchy requires more energy.

Metaphors of different types act through existing channels and sometimes build new connections. In a hierarchy of abstraction, similar flows operate. Metaphor can:

1. move easily from a higher to lower level of abstraction as in "La nature est un temple…" (Baudelaire. "Correspondences"). All of nature is likened as a unit to a particular temple with "living pillars."

Or Dickinson's "Hope is the thing with feathers." Metaphor generally scales from higher to lower levels of abstraction facilitating transport. The move toward specific references conserves energy due to the ease with which thoughts travel from general to particular. In some cases, this energy flows too easily engendering corruption. Generating multiple paths, operating with constraints, and gating of flows are poetic practices.

2. move horizontally on the same level where two humanist abstractions, both poetic concepts, are compared as in Hejinian's equation: "What is love but imagination."[10] And in another comparison of equal generalizations in "seeds" like "corpses" in Shelley's "Ode to the West Wind." This transit is also easy and stimulating because of the low energy required to move horizontally within any level of hierarchy.
3. move indirectly and with difficulty to a higher level of abstraction (larger scale) as in Donne's "She's all states, and all princes" ("The Sun Rising"). Here the connections between the source and target are fewer and more difficult to access, requiring the reader to select appropriate flows. Only a few of the many components of the power of states and princes are invoked. Power over the lover (who "nothing else is") establishes an obsessive singular identity that like other singular identities mistakes itself in its squalor as unitary and alone because, of course, the lover is also the unacknowledged writer.

The reader is required to travel upward to a higher level of abstraction which requires more energy to understand like the problem of the brick layer. The reader also has to move from the domain of the lover which they easily feel to that of politics which they must think about to understand. This upward motion may mean no more than Donne's love makes his mistress as important as a prince either in his mind as a submissive lover or the deluded mistress opinion of herself or both if they share the authority fantasy. Lots of ambiguity arises when moving upward in the hierarchy. For the attentive reader, translation to a higher level of abstraction has environmental overtones in highlighting the similarity of different conceptual fields and scalability in the biosphere. This complexity and difficulty of the metaphor also shows why metaphysical poets fall out of favor when readers demand visceral and immediate impact from poetry.

Metaphor moves in six directions—
between ourselves, societies, and surroundings—
through scales of abstraction
and within each of the three.
Processes act like things
under certain conditions
and things act like processes. (Barad)
Things combine in multiple forms
as nodes and as channels between them.
As node both perceived and interoperating,
as channel carrying language
and in language metaphorically
things scale in three directions.

(The limitations of singular identity appear transparent once we look at identity through this hierarchy.)

Metaphor looked at this way operates more like an ecology in the biosphere. The metaphor is not merely conceptual (Lakoff). Physical similarities and interactions create and operate metaphorical processes within/through bodies, between bodies, and in combinations. The senses perceive similarities and build models or schema to help identify the same phenomenon next time.

Once children learn to identify a horse, as Tenenbaum proposes, they compare the image in front of them to the schema of horses built in the brain. A parent must point to a horse only once or twice before the child rarely mistakes a horse for a cow. The schema is easily built since the brain is an efficient inference machine.[11] It is no accident that the model of ecological connectivity and the metaphor of movement through a hierarchy can be spoken about together and play together in these language games. They act isomorphically across conceptual and physical boundaries that we can understand as no longer binary. Language reconstitutes and affirms the world in the brain through these schemas because it evolved reflected, refracted/bent, and diffracted/spread through the world as a model. The twisted light of perception flows throughout the hierarchy which is why we need models. The mental model is visible in the world at the bodily, social, and ecosystem levels.

The environmental model intends the reader to extend socio-centric concepts of contemporary criticism to include the surrounding world where metaphor builds through the organismic binary as in poet erica kaufman's metaphor "material troops of fortified islands."[12] We are not helpless cogs but change the world as we reorient our ways of using language as well as building our surroundings. This is true for our identities as well as our activities focused on family, job, community, and power over ourselves and others.

In this sense, the environmental model draws a map (another physical/conceptual metaphor) of metaphor. As Lakoff says in his essay "The Contemporary Theory of Metaphor," "The mapping is the set of correspondences [Baudelaire, maybe]. Thus, whenever I refer to a metaphor by a mnemonic like LOVE IS A JOURNEY, I will be referring to such a set of correspondences."[13]

Correspondences in the environmental model, whether Baudelaire's or Lakoff's (to the extent they are different), use ecological channels throughout their interactions. Metaphors are endemic/inherent/"natural" to the world we live in as part of our surroundings, as social constructs, and as built in the brain. They effect the evolution of our thoughts, bodies, and societies and include climate change feedback to the source ecosystems that themselves, metaphorically and physically, may be mental. Metaphors of travel, competition, conquest, and others physically impact the biosphere. The biosphere in turn operates selective evolution through species and societies. It operates with similar methods—force majeure, interaction with and control of surroundings, chance, rationality—and processes to build the taxonomy of metaphor into society and individual brains. Evolutionists do not call it metaphor but one thing stands for another everywhere we look in both biological homology and analogy.

Reddy describes how the brain manufactures isomorphism/similarity/likeness to facilitate communication.

> Language seems rather to help one person to construct out of his own stock of mental stuff something like a replica, or copy, of someone else's thoughts....[14]

This replication process is true for other inputs to the mind as well as language. The inference machine of the brain ensures a tropism toward the frame one is using and, in Erving Goffman's terms, general frames propel that tendency.[15] Reddy's formulation supports the environmental model up to this point where "English…can bias thought process toward this framework…"[16]

The speaker is reassembled/rebuilt by the listener after translating through the channels of communication and with a shared process between source and target. This accounts for both misunderstanding and approximation of understanding. Just when Reddy is on the verge of expanding his framework to his surroundings, he insists on staying within a language model to critique the conduit metaphor instead of moving to the ecosystem metaphor where his other metaphors operate by reference. In the environmental model of metaphoric and physical networks, in order to know, to communicate, and to be, people project sets of satisfactory metaphors channeled through conduits transiting eyes, mouths, ears, skin, and brains. The metaphoric network operates similarly to the physical ones because the metaphoric ability of the mind builds thoughts that look similar to their physical context, that is, context spawns the nominal scaling that metaphor can bridge.

When we use language, both words and their order include and adapt to the processes of communication, construction, repetition/replication, and performance. All four exist through interpretation and all four act together. The network and ideas conceptualized by and processed through the parts of speech and their grammars of construction are well classified if fuzzy and thick at the edges like most categories.[17] The network, language about it, and relevant ideas operate through a hierarchy that looks more than a little like Linnaeus' taxonomy. The hierarchy is presented as a sentence.

Level 0: Language operations include
Level-1: Concept, meaning, and perception that trigger
Level-2: Memory and interaction with surroundings both as storage and effects of surroundings. (I call your name and the birds fly away. Sound waves carry through the air on channels dispersed by the wind.)

Within another hierarchy, language is not only a conduit for meaning. Like other ecological processes, language develops through channels between parts of the brain and interacts with external channels both through processing and in the way the world acts as a storage device for the brain. As a metaphor, language carried over channels in envelopes is addressed to recipients specifically and generally such as you reading this book. I speak to you about networks; my messages are carried over the network; you reassemble my message in your brain's network. None of us understand quite the same thing, but most understand similarly, isomorphically. You may understand well, somewhat, or poorly depending on your reconstruction of my linguistic performance, and Reddy is right about that, too. My only concern is his incautious exclusion of the metaphor of language as conduit, an organismic foible reiterated in social structures his organism built and accepted. I'm sure that at several levels, I, too, am sharing a socially structured segregation of thought, and a protective mental process.

Initially, Reddy is concerned about "*how that problem* [of communication] *presents itself to us*." Initially, he says, communication is "determined by the structures of language itself" and that "merely by opening our mouths and speaking English we can be drawn into a very real and serious frame conflict."[18] Reddy suffers from trying to prove one thing and disprove another when both have elements of truth depending on the frame of which he speaks. The ubiquity of conduit metaphors means that it's one of the useful ways to communicate about communication. There are too many of them to be completely discounted. And as Reddy points out, up to "70%" of metaphors refer directly or indirectly to conduits. The operational frames that the matter multiply.

The multiplicity of connectors requires a model that I have already said can operationally look like a network and then ontologically adhere to a Venn diagram. Trying to say something that is ubiquitous is not the case fails the sniff test. Multiple solutions to complex problems make more sense that the either/or binary Reddy is anxious to prove. The subtlety of Reddy's argument runs aground in that he fails to address the ecosystem but only conceptual frameworks. Of course, if you're stuck in the mud of certain logics, you will need to get pulled out by increasingly arduous methods. The message traveling across the network is only the data and does not address the metadata that determines that your message to me comes to the right place to be understood. We need to retrain our brains to multiple possibilities.

Language performs, communicates, and constructs one or more versions of a thought and its audience, the reader in this case. Metaphoric language gives rise to other metaphors that Reddy worries might not be accurate enough, but what can you expect of a metaphoric stack in terms of accuracy? As we have seen since Reddy's article that the too perfect communication of English through social media may well cause more harm than the problems it solves. We must appreciate precise communication as in mathematics, conceptual communication about events as in social sciences, and probabilistic communication through senses among many others.

Approximation is the limit of accuracy both in language and in environment, although we improve approximation as we get closer to specific predicted events like a rainstorm and a viral pandemic. The perceived events unfold and are coded in memory showing how the metaphors are indissolubly linked to physical entities even if through another mediating entity like an idea.

These processes are physical in the sense that every metaphor is generated in the mind by a set of relationships between parts of the brain as in Graziano's "attention schema" where generations of paying attention have developed a model or schema of attention in the brain. The faculty of attention that is aware of smell and the model of attention exist in separate parts of the brain.[19] These schemas interact across and build hierarchies that take us further away from physical reality into the conceptual that only some of you will agree is also physical, but which we all agree that we know something about what we mean. The phrase "I know nothing" can be wrong, disingenuous, and deceptive all at once while at the same time expressing genuine despair. Social agreement and disagreement generate opinion and resistance.

It is not that "language functions like a conduit," even though it is so in metaphor, but rather that language processes through connections that also have linguistic structures and performance properties. In that way meaning is processed through language structure as Reddy himself asserts. It is generated, translated, and then spoken to the air as sound waves, transmitted, and then reconstructed in the mind of the reader and turned into a set of responses: levels of agreement, echoes with isomorphs of the message (metaphors), emotions, and physical responses like increased heart rate. Language is processed through physical conduits *and* language can be described as a metaphor and used with metaphors about physical networking. Isomorphs work like that. Changing frames and reciprocal frames are already processes of connectors or a node at the intersections of connectors.

One important caution regarding the binary, conceptual/physical, that Reddy and Lakoff use to persuade, it's not a stable binary. Ecological channels include a range of physical and conceptual linkages without always a clear border between them. Even thoughts in the brain have location, charge, and direction. The spectrum of connections includes and is not limited to:

- insulated channels like electrical cables
- encapsulated vessels like capillaries that leak, absorb, and allow passage of nutrients through them
- delicately wrapped neurons that pass charge across synapses that operate by a combined chemical (changes in ionization) and electrical process
- semi-permeable membrane that privilege access in multiple stages to certain chemicals across cell walls such as cancer pathways, epigenetic changes in individual cells, and cultural influences through which groups evolve
- passwords typed in and then translated and interpreted to open switches allowing access to systems and specific data
- light as interpreted by eyes to trigger concepts that only occur with certain types of excitations of the brain
- ideas that are usually linked within models to other ideas
- concepts that have boundaries only through false consciousness driven by social activity
- concepts isolated for heuristic purposes that are then linked to other concepts for more complete description and practical action of those concepts on matter
- the newer metaphysical usages of the word matter that include physical and conceptual components
- concepts that appear in the mind after two or more parts of the brain connect electrically through neurons that link through writing to social engagements between citizens. Foreshortening of this set of linkages can lead to excessive violence
- confounding variables bridge the binary that swings back and forth across the chasm between things and thoughts about them.

The combinatory propensity of nouns also implies that no thing, physical or otherwise, operates by itself. All concepts and physical presences have their isomorphs reflected and refracted throughout the biosphere. All these actualities complicate the binary beyond usefulness.

Now I return to the sentence ("Language operations include concept, meaning, and perceptions that trigger memory and interaction with surroundings both as storage and effects of surroundings.") as it operates in the three-level hierarchy outlined above. Lakoff uses the phrase "an ontological mapping across separate domains," implying that metaphor operates at several semantic and conceptual levels of the three-level hierarchy. Since meaning operates in the brain on internal signals as well as between the brain, its body(s), its social fabric(s), and its habitats(s), the physical process of mapping can be said to take place both inside the brain from lobe to lobe and through channels between the brain and outside to include other parts of the body, because as we know thoughts affect bodily functions. Think how seeing the train approaching stabilizes and directs your movements.

Once beyond the body the interactions include a social process of mapping and a process that engages various components of the habitat. In this way, cognition takes place through connectors to outside as well as within the brain, an environmental idea based on the work of Mark Rowlands.[20] The further away the less impact in most cases, but doesn't seeing videos about microplastics in the Sargasso Sea extend your thought there? If these connections have both material and conceptual components, where do you store a connection between yourself and the sea? Tenenbaum might say that you build it from your connection schema.

When the metaphorical concept moves in a downward direction in the hierarchy, that is, to greater levels of detail and lower scale, mapping and comprehension of the map are easier. As mapping moves upward in the hierarchy from specific to general, traffic is slower and more indirect as in the Donne metaphor discussed above "she's all states".

The environmental model goes further toward materializing metaphor in that thoughts and actions travel across these channels within and beyond the body, within and beyond the social to surroundings—a map that extends throughout human and non-human species as in Lakoff's, "The language is secondary. The mapping is primary."[21] The map of the networks enables understanding. It labels thing-like nodes and follows the channels of both communication and event sequences. Whenever I assess my identity and look at myself in comparative (metaphor as one kind of comparison) ways, the mapping process captures a portion of my surroundings, tracing social and individual paths via senses that notice sound, light, smell, touch, and meaning that I focus on configuring material routes. What Lakoff calls maps, the environmental model extends to a network mapping process that engages specific subnets of brain, body, society, and surroundings in the hierarchical manner outlined above through its indigenous notion of traffic and travel. So, here, metaphor materializes a tangent to its own application.

That metaphor that results from physical interactions has precedence in the experimental work of Graziano and previously Frederic C. Bartlett. In his essay "Imagery in Thought," Bartlett shows that in the mind "images are particularly liable to arise when any slight check occurs…"[22] He means that when the logic of a process mapped in the brain—the connection map, for example, between sense input and its schema—fails to make a scheduled connection, we project an image to bridge the gap and continue thinking. Only after clearing up the confusion does the network continue along known paths and the "logic" of mind returns.

Framed another way, when the expected logic of a process breaks down, images arise that provoke expressive responses such as metaphor around individual fears, social unrest, and ecosystem disruption. Climate change is such a collapse of both local and global processes generating images on the largest scale in human history since the ice age. Many people including myself find these metaphors of climate change almost inconceivable and certainly intolerable leading to the misconception that humans cannot conceive of the scale of the problem. Sometimes, poets despair. The scale of climate change is big and difficult but not inconceivable.

Metaphor channeled through the routes on the physical network in the brain extends performances to the rest of the body, society, and the world. Interactions and the isomorphism of different levels from thought and blood flow to rivers and supply chains to the influences of theme and style between poets all operate similarly. Metaphors are more than terminology and figures of speech. They are physical correspondences that travel ecological networks carrying meaning and ontology, connecting A to B, love to bees, and so on.

Lakoff's "open-ended class of potential correspondences" refers to material connections between parts of the brain, even if they are not so easy to find and track. Yossi Arzouan et al.:

> The present study demonstrates that RH [right hemisphere] mechanisms are necessary, but not sufficient, for understanding metaphoric expressions. Both hemispheres work in concert in a complex dynamical pattern during literal and figurative language comprehension.[23]

To mention poetic innovations, in comparing two-word metaphors, such as slippery slope, from poetry texts, the same researchers found that different parts of the brain are engaged when processing novel metaphors than other more "literal" expressions. In studying metaphors in Mandarin, Kathleen Ahrens et al. found that:

> conventional metaphors differ from the literal condition with a slight amount of increased activation in the right inferior temporal gyrus.[24]

Through connections in the brain, metaphor processes differently than other language structures.[25]

There is every reason to realize now that the physical connections in the brain discovered by physics, chemistry, and biology also operate mental networks through poetry and psychology where different points of view are specific mappings of the brain. Speaking hierarchically then, the brain connecting, both within itself and upward from level 0, operates subnets called mind, heart, soul, and perception processed from the senses and other functions on level 1. Processing, realization, performance, and communication are all different mappings across the network as poetry impacts the reader/hearer.

The lists of true, somewhat true, somewhat false, and incorrect statements about conduits that Reddy points out are good examples of how to debunk fruitless and sentimental attachments that do not function in the world but appear in language and perception. Like the glaze on a doughnut, they stimulate the system while causing it to break down more quickly. They are by inference another climate change problem addressing concerns about fossil fuel in development projects. Narrowing the focus to problems with fossil fuels is easily digestible but doesn't provide much sustenance in rethinking our relationship with our surroundings. Monological thinking recapitulates the organismic binary. Yet climate change must be remediated step by step and each step toward the ideal must be real.

Poetry benefits from using hypothetical and abductive logic that resists single-value solutions and binaries to validate the point about the physicality of metaphor:

1. We know that the planet is warming
2. We are less certain as to the cause, although we strongly suspect greenhouse gas emissions from the six human industries.

3. We are more uncertain as to the best solution partly because
(a) many solutions present themselves and specific interest groups often promote solutions that improve their position rather than the considered, long term, planetary, or even narrower human interest
(b) we don't know what domains—psychological, social, and habitat—will more readily respond to alternative processes
(c) there remain concerns about the solutions themselves because of the downstream negative effects of implementing those solutions both in terms of risk homeostasis, ecosystem receptivity, and surprise.

How do we address such uncertainty? Certainly not with absolute, moral, or binary thinking, although picking one by gut feel/instinct/chance and going ahead remains attractive to the organismic binary. It's surely faster. But I'd point out that we've been trying that approach for millennia. And in some ways, there is little alternative, but as an alternative to gut/thought conflicts:

As each node in a process
becomes clearer, transparent,
approaching through statistical likelihood,
trial and error, poetry,
with mere additional confidence,
moves to the next;
choice and chance operate
simultaneously in cocktails.

Returning to the materiality of metaphor, a range of possible solutions mixing the purely literary, the linguistic, the conceptual, the social, the biological, and the ecosystemic appear. Several levels of the biosphere's hierarchy help confirm how metaphor works. Through isomorphic (scalable) connectivity metaphor may be:

1. Processed as a literary trope where one thing stands for another as in Baudelaire's phrase quoted above: "La Nature est un temple…" Writers and everyday speakers often use nature as a metaphor to refer to non-human activity. Remember, these literary metaphors operate in three directions through different levels of the hierarchy. This statement refers to nature as the whole physical world as opposed

to the part of the world humans need to protect and sustain against industrial incursions and as opposed to my preference for the term as inherent characteristics.

2. Processed between two or more parts of the brain as in Graziano's attention schema and Lakoff/Reddy's notion of metaphor as conceptualization. Lakoff suggests a binary:

> Abstract reasoning is a special case of imaged-based reasoning. Image-based reasoning is fundamental and abstract reasoning is image-based reasoning under metaphorical projections to abstract domains.[26]

The opposite, where images only occur when logic breaks down as in Bartlett above, may also be true. Environmentally, I'd suggest that both are possible options under different conditions. Conventional metaphors make sense to use across cultures because we use them daily and many are common across cultures. More is up, life is a journey, etc. (Reddy) These metaphors work in and between ecosystems. Purposes are destinations because we so frequently have to move to achieve our goal; the metaphor is physical and reminiscent, recollected in tranquility or spontaneously enacted. It's also coincidental, linguistic, conceptual, and an illusion: environmental in its diversity.

3. Fed back between the brain and other parts of the body where the brain develops a connection due to a particular configuration and status of the body. We feel depressed from excess stomach acidity and report sadness. Sex excites the mind and vice versa. Such organismic events can be traced to psychological, bodily, evolutionary, social, and environmental sources.
4. A social network where one event triggers another along a chain of interaction. Social processes also include reasoning through successive approximation as in legislative changes toward shared power, freedom of speech, desegregation, reproductive justice, and finding the best methods of limiting climate change.
5. Found in interpenetrated ecosystems. The biosphere operates through processes similar to metaphor where one thing frequently looks like another. Homologous features like wings on a bat and wings on a bird occur because of environmental constraints and not due to a close genetic relationship. Feet and flippers operate

> like each other in different media. A part can stand for another through camouflage where stick bugs disguise as twigs. Functional relationships also promote metaphor through complex, polymorphic identities. From other species' perspectives a tree is something other than a tree to itself: shade, food, home when the tree is all of those, some to itself and some to others.[27] (von Uexküll)
>
> A person also has many identities as pointed out in earlier chapters: some to themselves, some to their familiars, some to others in the social network, some to outlying social networks, and an environmental identity that may not support an individual's identity as it appears to them. Human use of metaphor as in the Tenenbaum article cited above develops in a learning process from these physical and conceptual transformations. Your identity is not only who you think you are since others both contribute to your view of your identity and think of you differently than you yourself think. You are they.

If the reader accepts that discourse is material, then metaphor is a physical connection between nodes: what goes on in the mind is substantial, natural, comparative, and hierarchical in the nested sense outlined above. Discourse, not only a material thing and process, also through reference, acts as a connector to other parts and frames of the world. Environmental metaphor establishes one or more connections between two or more things sharing some characteristics, that is, isomorphic things that allow some bleeding into other things through metaphor while retaining their own continuous, iterative identities. (Time is money. They are also themselves and can be other, too, like time is space.)

Organisms, groups, and their surroundings interact, communicate, construct, and perform through the material connections of a global network that reflects, refracts, and diffracts the others in that list. Humanist culture models reflection. Diffraction occurs as these relationships spread across our bodies, social experiences, and physical surroundings to produce a more diffuse but highly related set of surroundings and concentrated selves. Refraction occurs as we look at things and events through the lenses of ourselves that include variable sense organs, variable situations, variable access to diverse memories, and bodily conditions that change with input and the body's own cycles. Champika Epa Ranasinghe and Christian Kray (Fig. 1)[28]:

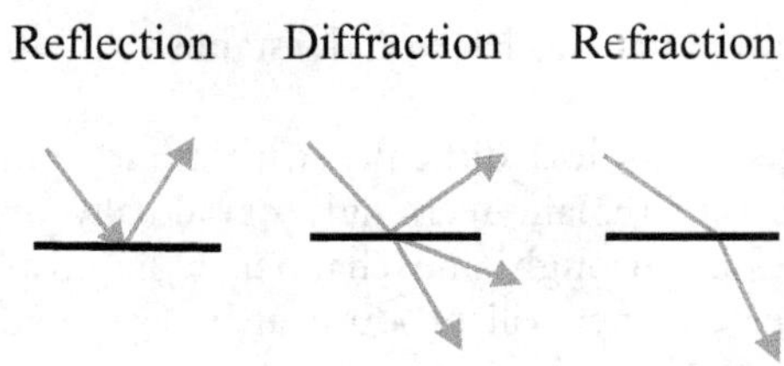

Fig. 1 Reflection, diffraction, refraction

Metaphoric comparisons and substitutions are generated by interactions in the brain and are also created in the relationships present in social and ecosystem settings. Writers get good at them with practice, implying that a physical connection can be established in a way that can improve the creation of novel metaphors. At the same time, everyone generates images through well-trod paths of memory, pop culture, and genre writing. Images arise in the brain commonly from intended constraints and links already established in mind, society, and habitats.

Reddy shows how all language is fundamentally metaphoric. Since the world is structured linguistically, how then is metaphor so different from the material world? Configurations and entanglements. Metaphor is a product of the material world. Material language processes material things with its matters which are words and concerns which connect. When we arrive at the edges of material constructs in electrical, conceptual, and other conditions that are less thing-like that transform back and forth between configurations easily, then language, society, and physics have some difficulty describing and even dealing with them. Entanglements abound as poets and poetasters misconceive who they are. As Armantrout says in her poem "Close":

The crowd is made of
little gods
and there is still
no heaven[29]

Churches, cults, and other institutions attempt to address them and the unsettled feelings they produce in material. As Barad and von Uexküll point out by different methods, things, too, have these multiple states, multiple identities, as in the diverse examples of sub-atomic particles and trees. The environmental model would establish this point for human identities as well.

Environmentalism has less difficulty with these transitional borderlands of domains that remain seemingly or actually immaterial. When a thought progresses through one channel, a linguistic connection is made, but another channel will produce an image; maybe many channels produce images. Abductive approaches to problems support solutions to anxiety, social alienation, and biosphere degradation, although the solutions are not always supportive of individual identity. Sometimes the situation supports the group. The biosphere created and is created from the network that evolved in the brain, that learned from the biosphere, creating mind, which is after all a "gift of nature."

1. All language is metaphoric
2. Metaphor processes through material networks
3. Metaphor occurs when the logic of language changes quickly
4. Then the flow through the brain changes in an identifiable and often obscure way
5. Metaphor bridges forms, times, and states
6. Through these difficulties, selves, a composite of identities, materialize.

Connections Below Form in Poetry & Biology

"Spare change? I can't even change my mind." *Conversation between two denizens of the Bowery overheard by the author in 1977.*

Environmental Identities

This chapter examines how identity develops similarly to syntax. First, I describe seven exemplary identity relations at different scales. Then, I present Michael Thompson's model of environmental identity as an alternative to Chaucerian humanism in *The Canterbury Tales* where one's job determines their identity, presaging scientific and capitalist specificity. Although people often apply identity to the whole organism, this chapter shows how identity is conditional and serial, built up in layers, and based on an individual's changing roles. Writers have multiple identities. We judge writers by their work alone and we also evaluate the writing based on non-writing criteria? Adjusting Whitman's poetics, individuals have multiple selves and songs. Language constructs the biosphere and human identity. Life in its excesses will continue to assert.

* * *

ONE: Diverse locations, cultural biases, and experiences build similarly shaped but separate identities step by step. As location changes, the cultural bias changes but retains its overall appeal to diverse identities. For example, Russian views on evolution differ from Darwin's. I associate the Russian position primarily with Kropotkin but many Russians scientists shared his view. Daniel P. Todes:

> Proponents of the 'strong' variant, however, went far beyond Darwin in their acceptance of four basic tenets: (1) The central aspect of the struggle

J. Sherry, *Selfie*, https://doi.org/10.1007/978-981-19-4870-1_15

> for existence is the organism's struggle with physical circumstances, (or less frequently with members of other species); (2) organisms join forces to wage this struggle more effectively, and such mutual aid is favored by natural selection; (3) since cooperation, not competition, dominates interspecific relations, Darwin's Malthusian characterization of those relations is false...[1]

Russian surroundings and cold climate also encouraged a different view than Darwin's entangled bank:

> Middendorf commented that 'this region's very scarcity of a variety of animal forms facilitates a better understanding of the general laws of life'.[2]

Today's similarities in laboratory biology build similar cultural biases for scientists. Globally, scientists find it relatively easy to exchange ideas except as restricted by politics, oh, and competition as well.

In poetry, we see how versification, metrical techniques, and forms vary from language to language, but poets' identities are similar and reproduced even across cultures and climates such as the anti-social hermit and the diplomat poet. How do lifestyles and locations affect the poetry's form and theme?

TWO: The environmental model's approach may be viewed by certain people as uniquely Western and capitalist, oriented to privileged white culture, or as heteronormative and ableist. If so, each geography, nation, and social group impels redefining and proclaiming separate sets of possible, unlikely, and out-of-bounds' behavioral solutions in modeling environmental explanations and remediation, and in writing poetry. Even at the small group level, unique cultural solutions to matters of language and habitat abound. These differences among individuals, groups, and locations certainly suggest multiple views of our surroundings and climate change solutions. There is no single solution and many possible paths to success. On the other hand, Deleuze: "difference immediately reunites and articulates that which it distinguishes, ..."[3] Practically, by putting people together or juxtaposing ideas from different disciplines, desegregation if you will, their differences and similarities both become apparent and less important.

The human ability to infer binds with morphology, as in the shapes of both organisms and thoughts, to form an environmental unit (called elsewhere an assembly and an apparatus) that propels ideas into categories to shorten the path to comprehension. Categorical thinking makes sense to many species and is not an arbitrary idea drawn in some mysterious way from the Cartesian mind isolated from the world but appears throughout the biosphere. According to Deleuze, these.

> ideas have a perfectly legitimate 'regulative' function in that they constitute true problems or pose well-founded problems. That is why 'regulative' means 'problematic'. Ideas are themselves problematic and problematizing.[4]

The combinative identity of noun-things described above gives them independence as individuals and connectivity in groups. Deleuze:

> Kant likes to say that problematic ideas are both objective and undetermined. The undetermined is not a simple imperfection in our knowledge or a lack in the object, it is a perfectly positive, objective structure which acts as a focus or horizon within perception. In effect, the undetermined object, or object as it exists in the Idea, allows us to represent other objects (those of experiences) which it endows with a maximum of systematic unity.[5]

The thing and by extension the self are both indeterminate and objective supporting traffic back and forth through borderland connections between and within ecosystems, social structures, and individual identities. (Taking my selfie locks in one instance of self at one time in one place). This problematic and plural solution model for climate change scales up and down to ideas about self, society, and surroundings that themselves must be viewed as problematic, that is, questionable, and therefore multiple and rangy.

How similar Deleuze seems to Bohr's indeterminacy through Barad although without the physical model.[6] How similar, according to Talcott Parsons, to Freud's connection between individual psychology and social structure in *Civilization and Its Discontents*.[7] How similar to modern poetry in spite of the narcissism of poets. These formulations are not created, as Deleuze says, but rather operate "within perception" to related (isomorphic) constructions within scalable components of the biosphere.

THREE: Looking at how I identify poetry as an editor, I realize that I test submissions to verify whether the connections they build between words, phrases, and ideas are problematic, effective, untried, politically perspicacious, and even able to be generalized. I tend to reject those that are largely reflexive, those that seem to avoid or ignore connections that scale. My test criteria will usually be less rigorous when reading a poet whom I already know or who I expect to operate within a known set of limits like a theme, a method, a school, or any identity group. I also test whether I can comprehend what the writer is trying to accomplish and how well they do it. There are countless details that I also consider, consciously or not, including when during the day I'm reading, the weather, and what else is going on in my life at the moment. If I'm flush with manuscripts, I read differently than when I need a book to publish in the current season.

FOUR: Corporate leaders' views of the planet also change with what they know about their surroundings, what's happening around them, and how they are mapped, i.e., Meriwether Lewis and William Clark, Ibn Battuta, and Zenh He's voyages. Notice how identity patterns changed when Europeans found America instead of India. Almut Höfert: The Turkish "menace" of sixteenth-century Europe was less threatening to the identity of Europeans once they themselves were conquerors instead of threatened by the Turks even though the Ottomans threatened the gates of Vienna.[8] Notice how perspectives changed when the photo "earthrise" was shown to the people of the world. But thought patterns do not change in one direction. Some see the planet as a unit to be exploited, some see its beauty, some see its unity, and some see threats to their local control. Leadership's judgments alone, even with the best intentions, will not get it right every time. Not all explanations and solutions work, some are inaccurate and some disguise other intentions. Some top-down solutions from leaders are no more than tactics to maintain power.

FIVE: The connections established between individual and group identity can best be examined in detail to avoid superstitious beliefs. US southern border solutions, for example, can be understood from conditions in the migrants' country of origin, from the different identities of the travelers themselves, and identities in the US border populations at the other end of the channel.

Human societies have devised creative ways to slow inevitable change in order to retain institutional identity. Poetry has books and schools to retain historical trends, but institutions promoting change in poetry are less effective than small groups of writers exchanging ideas to build new poetry. These groups act together overtly and consciously in creating tendencies in poetry such as the development of the tenets of language writing by the epistolary exchanges between Bernstein, Silliman, and Andrews.[9] Editors and critics also gather group advances through publishing poets together and writing about them as a group. The identity of poets changes through small groups as much as with individual creativity.

SIX: Domestic identities between Gertrude Stein and Alice B. Toklas explore multiple possibilities of meaning for Stein's literary identity by showing how some connections are traveled differently than intended:

> I was overcome with remorse. It was my fault that my wife did not have a cow. This sentence they cannot use.[10]

Stein's sentences open up several interesting options, some pragmatic, some comedic, and some esthetic. The wife (Toklas, I assume) cannot have a cow perhaps in the sense of not having an orgasm through Stein's ministrations, but connections in the other direction from Toklas to Stein work in the sense of communicating distress at Stein's failure. In sentence two, Stein feels remorse through internal channels. The first two sentences thereby form a unit of meaning that lead us to understand the domestic narrative. Sentences one and two show multidirectional connections between the two lovers where in one direction intention fails and the reverse direction carries multiple messages to Stein that cause her to feel remorse. Then Stein tries an unusual cognitive jump between sentences two and three by showing a different type of connection than between sentences one and two. Between sentences two and three, I read a change in scale from sentences one and two, what literary folk might call a meta-reference to the first two that put making love and writing together.

SEVEN: Further complicating the characteristics of identities that might slow climate change, explore biomimicry, and other solutions derived from existing biosphere processes favor completeness over efficiency while human solutions tend to seek efficiency and identities tend toward specialization, especially in a humanist-capitalist framework. Biologically, an organism has to reproduce, not establish perfection.[11] Through sexual selection, people progress toward ideals and more attractive esthetics that optimize an individual's chances for reproducing, improve group regulation and control of individuals, parental discipline that reduces confronting the child, and ego gratification. Sexual selection may drive resistance to strategic climate change solutions that contain apparently dysfunctional or inefficient components such as components that operate in multiple ways and components that function all the time but are only needed occasionally. Can the drive for efficiency in both economics and politics accept such cultural divergence?

Stoical social planners can look forward to these inefficient practices to provide sustainable solutions. From the point of view of an individual's identity, this means recognizing that single identities apply only under certain conditions and then change. Infallibility and consistency are unlikely for long even though the organismic binary twists reality so that I try to appear right every time and often defend strange and annoying behavior. Stories of this type abound in my biography and perhaps in yours, connected reader.

When someone clarifies their difference through a specific action like:

- walking while Black
- coming out
- realizing religious abrasion
- acquiring some class privilege
- acknowledging that your culture has committed heinous atrocities
- noticing that some people think they are better than you
- noticing that you think you are better than some other people,

identity gaps and discontinuities begin to form through inefficient use of channels. Some gaps are discrepancies, some chasmic, and some, as a friend once said, are chasmic discrepancies. The linkages bridging these widening gaps change their characteristics with distance, time, and condition:

- A relationship between two friendly poets begins to be tinged by racial and professional overtones and later assumes privilege in an unusual way as a woman begins to leverage her academic position.
- When my salary goes up, I wait for you to call me.
- When my new book comes out, I no longer acknowledge you at parties.
- Threatened masculinity results in an affinity for specific kinds of ritual violence in poems and movies.
- When the Jews gained a nation, they improved their identity for themselves and for the world, but anti-Semitism persists.

- When Barack Obama was elected president more Black people were put at risk in their own communities by white nationalists even as they gain stature in their own eyes and in the nation as a whole.

Unintentional actions have different characteristics of connection to that which is acted upon than willed actions, but today many conservatives and progressives are as unforgiving of a slip of the tongue or reversion to habit as of an overt act because of what Freud assumed about psychological scaling, how we were trained as children to behave, and what we can exploit to attack the perceived opposition. The common idea of scaling erases some boundaries of category and classification.

In another dimension, people with different jobs and roles see their surroundings differently. For example, ecologists see available resources as limited while entrepreneurs see the world as a cornucopia controlled by skill. Children see that cornucopia as freely available; unhappy people think things are not in their control and some people don't want to engage except in extreme conditions where little ambiguity remains. Recognizing these different perspectives—diversity if you will—I was attracted to Michael Thompson's environmental view of society in "Man and Nature as a Single but Complex System"[12] and subsequent writing.

The generic roles, as defined by Thompson, negotiate diverse views with other people in order to achieve sustainability—a social model—and with their surroundings. Not only do people negotiate overall perspectives on their surroundings, they also vary their views depending on the different roles they take as pointed out above. Thompson's views of nature are not defined by an individual identity as a whole but by the role of the individual in different situations. My point of view changes depending on whether I am a writer at my desk, an old man watching a young woman jogging, a voter in an election booth, a driver standing in line at the DMV, or a tourist strolling through a museum or a nature preserve. Some combinations of nature and role are operative while some as pointed out don't function too well, like Stein's cow fault.

Social being/voice	*View of nature*	*Reassurance/group relationship*
Hierarchical	Isomorphic nature	Government, law, critic, formalist, experimentalist as Oulipo: *seeking control through conditional grammars*. Georges Perec, Oskar Pastior
Egalitarian	Accountable nature	Activism, sects, political poet as Russian (ca. 1917) and Chinese (ca. 1911 & 1949) revolutionary poets: *Being part of the group as support for its politics*. Vladimir Mayakovski, Velemir Khlebnikov, Guo Lusheng, Bei Dao, H. T. Tsiang
Individualist	Skill-controlled cornucopia	Entrepreneur, markets, avant-gardist as language writing: *reassurance from group poetry and critical sources that you are in control of your style and expecting success from individual skills*. Lyn Hejinian, Charles Bernstein, Ron Silliman
Fatalist	Chance controlled	Ineffectual, victim, purist, romantic as ecopoet: *mind is reassured by being in nature of privileged access limited resources*. John Muir, Gary Snyder, literary critic Timothy Morton, some identify as workers like Jack Collom
Autonomous	Freely available cornucopia	Hermit, hobo, innovative poet as writers rejecting their group affiliation: *seeking order in randomness*. Average medical hero, Francois Villon, Jack Kerouac, Han Shan

These cells in the matrix (and individuals in them) are not fixed positions but dynamic relationships. Individuals extend and groups move from one to the other depending on role and also condition, location, and activity within a role. Some writers change location during their careers. Ashbery's *Some Trees* and *Three Poems* belong in the hierarchical category along with Oulipo. Later, for me starting with *Self Portrait in a Convex Mirror*, Ashbery's poetry changes as exemplified by his remark to me at the Local/Locale Bar where we were drinking with John Yau, "James, do I have to write experimental work my whole life?" Then in *Flowchart* and *Girls on the Run* his skill at building an esthetics around referentiality moved his social being toward fatalism and in some ways toward language-centered individualism and environmentalism. Ashbery as a poet is connected to but also different than Ashbery as a formative gay writer, activist in his profession, and critic playing with language. And I foresee problems arising from too cleanly separating roles. As Clark Coolidge wrote "Can they afford to scant what's not walled in?".[13]

Part of the reason to expect unstable relationships in the matrix and in any analysis of environment remains the range and scale of human interaction, wider than any one culture anticipates in seeking to control and celebrate participant's actions to throw poet's a bone. Important events that support this perspective come both from outside the culture like the Soviet Union's impact on America that for decades kept corporate capital in check and within the culture such as the frontier spirit in the US that made it such a fertile ground for libertarian conservatism. Literary influences from both inside and outside any language are numerous, including multiple interactions between French and English poets such as Baudelaire with Poe and influences from inside such as Kenneth Rexroth's influence on the Beat poets. Groups evolve from external influence and internal influence: inter-intra-action requires connectors.

Can group cultures and individual faculties evolve together to simultaneously keep group, surroundings, and individual identity functioning together? Threats loom so large in our fields and some point to the end of ecosystems support for culture as we know it and the necessity for species adaptation.[14] (Jem Bendell: Note this paper was hotly disputed by mainstream experts).

Which of these points of view is correct? Regarding fruit, steel, sleep, friendship, and poetry, the question of whether resources are in abundance, limited by volume, limited by skills, out of our control, or mostly irrelevant to what's important depends on who you are *and* what you're doing there *and environmentally,* where you are.

Reading this chart, you won't find yourself in any one place, but you might try. I know that I did over and over until I realized, with difficulty, that I change my cell as my role changes. I tried twisting myself into various shapes to conform to values I thought were more useful, moral, attractive, or essential. Then I stopped using the *or* of the organismic binary and realized *and* represented a more effective reading of this chart.

Marshall McLuhan's "the medium is the message" is a metaphoric alignment of subject and object intended as social critique identifying emerging digital culture. From the environmental perspective, the medium embeds itself in the message becoming the subject in a linked phenomenon. The medium includes the message as object, metadata in the lingo of the security state, and is subject to correlation with other metadata to produce actionable information. Efforts to create or find a single, essential identity fall prey to treating the message as an ideal, sacrosanct content. Although I cannot stop seeing my selfie and warding off threats, I can find myself more completely by including a view of my surroundings. What's behind me when I look in the mirror? What is my role?

These identities—writer, publisher, father, citizen—I speak from, about, and act within have placed language writing in the cell related to a skill-controlled cornucopia view of resources, because the central difference between language writing and prior American poetry emphasizes and intensifies the political meanings of technical skills of poetry. In addition to those components, language writers speak thematically in a public way about equality and egalitarian principles that approach poetry as words on the page, the equals sign of the magazine L = A = N = G = U = A = G = E. Third, the critical bent of many language-writer texts, with their wide acceptance within the academy, points to how language writing fits into the social schema. Language writers in their role as poets focus on skill. They refer to socialist and egalitarian themes, like revolutionary poets of the early twentieth century, but through technique. They have found jobs in the same universities as poets like Rosemarie Waldrop who use a less politicized esthetic abstraction and universities dominated by poets following New Critical theories. This university ecosystem allows for several species of writing to exist side by side, but competing for the same jobs and prizes while marginally overlapping in their publishers.

Environmental Autobiography

The chapter "Environmental Autobiography" presents four exemplary autobiographies of four, successful, professional women—Amina Tawasil (anthropologist born in the Philippines), Dorine van Norren (Dutch diplomat and environmentalist), Rae Armantrout (language writer), and Lyn Hejinian (language writer). I chose these four for some diversity and some overlapping of their situations. I also chose them because all four are at similar points in successful careers, but from different backgrounds and parts of the world. The similarities and differences show how assemblages dominate both identity and discourse.

* * *

Now let's move from building analytic structures to how particular people besides me look at their surroundings in four short environmental biographies. In 2019, I asked four successful, mid-and-late career professional women to write autobiographies from the perspective of availability of resources. I didn't stipulate much else except they all knew that I was writing another book about environment. Each narrates a person coming from several different perspectives:

Amina Tawasil is an anthropologist at Columbia University. Her doctoral field work focused on the women seminarians who made policy for the women of Iran. She is a Muslim, a mother of three, and writes in the third person from her socio-political view of self:

J. Sherry, *Selfie*, https://doi.org/10.1007/978-981-19-4870-1_16

Take the case of a PhD anthropologist Amina Tawasil who spent part of her childhood among her father's tribe, the Tausug, who are indigenous to the islands in the Sulu Archipelago in the Celebes and Sulu Seas. The coastal clans constructed their homes on wooden stilts, and were historically documented as masterful sea navigators. This documentation was often in the context of carrying out piratical raids against the Spanish occupation from the 1500s to the mid-1900s. Because of the likelihood of typhoons destroying harvests and homes, they did not have a strong tradition of storing surplus from the harvest. The lack of attachment to material is transferred to an attachment to each other. The idea of self is more horizontal (collective) rather than vertical (individual). Their view of nature is that their lives and resources are transient. Only enough for immediate use is retained, thus, the name Tausug, which means 'people of the current'.

Her mother on the other hand came from the Tagalog region with a slightly more stable climate where much of the Philippine's export products are farmed and processed. There she learned about the value of building stable surplus against bad harvests. As an adult when visiting her homes, she has to take the attitude of the people she's living with. When living with her mother's people, she dressed conservatively, used proper manners and retained a demure façade. Her behavior among her father's people was improvisational in that the horizontal self, in other words, those around her mattered in order for her to matter. The idea was that 'what was good for two people was enough for four,' there was always enough for everyone at that very moment because the next moment there may be nothing at all.

Back in the US, although she sees herself fortunate for barely avoiding becoming an adjunct instructor, she is faced with the revolving door of non-tenured teaching. Disenfranchisement in the academic profession among thousands like her promotes the need to live in a suitcase, paycheck to paycheck, with a low salary compared to tenured faculty, and the instability of moving to another campus every two years. In America, her view of her surroundings had to change again although viewing nature as unreliable and providing limited resources remains her experience. Her relationship with her surroundings was not related so much to her self in some fixed way, but rather to her role and the cultural climate of the moment.

Dorine van Norren is a Dutch civil servant, researcher, writer, and artist. For me she represents the success of liberal European governance as a diplomat and writer about climate change and environment. Her transitions even in the abundance of post-war Europe appear to have been achieved through substantial expenditure of personal effort.

My parents came from diverse backgrounds. My mother grew up in a wealthy family where her father was a self-made businessman who could afford to buy one of the first cars in the neighborhood and take his family abroad on holidays. My father came from a worker's family where there was no money to buy toilet paper. He worked up to become a director of an institute and university professor. My parents understood each other in their Calvinist frugalness compounded with memories of shortage during the Second World War: nothing was to be wasted. Plastic sandwich bags would be washed and used again, while my mother would sigh in the 1970's: 'where do we go with all this plastic?' My parents were not poor, but we would never indulge in expensive clothes, toys or go out for dinner, let alone buy a tea at a café. Much cheaper to make a pot of tea at home. In terms of hobbies I had to choose, one hobby only. Of course, I wanted piano, horse riding and dancing lessons, but this was frowned upon as spoiling children. Life was regulated by the church with its Sunday gatherings and festive days. Indulgence was once a year at our family holiday, camping abroad, one month a year that I looked forward to the rest of the year.

I grew up in a rural village in the centre of Holland. My mother would get the eggs from the farmhouse and we would play in the maize fields. There was an abundance of fresh air, green fields and gardens full of flowers. For an adolescent, however, this was boring and as soon as I could I moved to the big city: Amsterdam, the town my mother had come from. An exhilarating time started where there were so many opportunities that I did not know how to choose, each one of them had my attention. No longer did I have to choose one hobby only. I decided on rowing, singing, piano and joining the board of a European students' association, which allowed me to travel across Europe. When the opportunity arose to study in South Africa, I was the first to jump.

My luck continued and after many exams, I joined the foreign service. Here I discovered a new truth: as a woman I was not equal to men. I chose to ignore it, surely it would blow over in time. Life was abundant, especially during my two postings abroad: Sri Lanka and Turkey. My horizon broadened ever so much and thirsty for more experiences, I fully engaged in meeting a wide variety of people and dealing with as many files as possible: from trade, to culture, human rights, political affairs, development projects and so on. I was provided with all the resources I needed to excel. In terms of natural resources Sri Lanka was exuberant and I never felt happier in my surroundings. Of course, travelling this much also had a downside, namely frequently exiting boyfriends who would want a career of their own. I had arrived, as one of my friends expressed. At an early age I lived in a villa, had servants, plenty of money to spend and friends in high places.

In Turkey the clouds appeared on the horizon with increasing worldwide tensions on the issue of Iraq (and the scramble for resources) and mounting disillusionment on my part. This was not what I had signed up for. Was I to serve these masters? Slowly my comfortable world started crumbling. I had arrived on the summit, only to conclude that I didn't like the view.

After four years I returned home, and fell ill. At the same time, it became ever more apparent to me that the achievements of females are viewed differently from those of males, an issue I could no longer ignore. I realized I needed a career change and this implied foregoing all the luxury that I had gotten used to. I moved back into my small community housing in Amsterdam, to the utter amazement and protest of my friends and family. City life also started getting under my skin. I missed the green and couldn't understand that children would grow up without smelling the flowers in the morning. It became a long walk through a dark tunnel which seemed never ending. I radically changed my views of the world and discovered spiritual wisdom, the only source, apart from the natural world, that healed my broken heart.

Slowly I carved out a new niche for myself. I asked myself what I could do, without having to go back to school or lose my income. I started a new hobby: writing. When the opportunity arose for a strategic advice position, I knew this was my new path. It enabled me to pour in my experience and meet new people, reorienting me towards science. I then gained one-year paid leave to do a PhD and though squeezed for time—a rare commodity—managed to finish in two. The journey brought new travels and worldviews—Bhutan, Ecuador and South Africa—this time broadening my concept of wellbeing. I discovered my artistic resources, designed the wheel of values, coupled it with my research, and opened my first exhibition in 2019. (www.ascleiden.nl/news/new-exhibition-wheel-values-africa-and-world) After 12 years I exited the tunnel. There I found a meadow of people with a positive outlook on the world, dreaming of new futures where abundance is shared and the natural world supreme.

Van Norren's thesis, *Development as a Service: A Happiness, Ubuntu and Buen Vivir Interdisciplinary View of the Sustainable Development Goals*, organizes some of the work she has done in environmentalism in the context of global diplomacy: (pure.uvt.nl/ws/portalfiles/portal/19859816/Van_Norren_Development_18_12_2017.pdf). Beyond her Northern European inheritance, Van Norren exemplifies how cultural change can impact climate change.

Next are the autobiographies of two successful poets from different backgrounds with similar career paths whose work became renowned through a combination of their writing skill and the rise to prominence of the language writing group. The differences of their perspectives illuminate the complex relationship between self, group, and surroundings.

Rae Armantrout's rise from the working class to international prominence seems almost coincidental to her view of herself, although I know her as talented, ambitious, and focused intently on her writing. As with the other poet writing a resource autobiography, she wrote more what she wanted to remember than what I asked her to write:

> I was born in the late 1940s and raised in a suburban housing tract called Allied Gardens in the city of San Diego, home to a large naval base. The houses in this development were intended to be purchased by returning veterans using V.A. loans, as the name 'Allied Gardens' awkwardly indicates—combining, as it does, the military character of the 'allies' with the presumed domestic tranquility of 'gardens.' The houses came in five models which repeated up and down the blocks. Ours was a 'Miramar,' which means sea view in Spanish, though the sea was nowhere in sight. We were about fifteen miles east of the beaches. It was the only home my parents ever owned, though my mother did enjoy touring the larger homes for sale in the new developments breaking ground around us. We would go on Sundays after church. My father rarely accompanied us.
>
> My father was a chief petty officer in the navy for most of his working life. (We never went to the beach because he said he saw enough of the ocean when he was on duty.) It strikes me now that 'chief' and 'petty' are at least as dissonant as 'allied' and 'gardens.' I wonder if my father felt that too. (Maybe his rank bothered him: I'll never know. He didn't talk much. I realize now that he probably suffered from undiagnosed PTSD.) My mother managed a candy store. She was one of the only women I knew who worked. My parents were children of the depression. Neither of them went to college. They were careful with money, allergic to debt. I was an only child. I lacked for nothing that I knew of—but there was a lot I didn't know. For instance, I didn't know that we were lower middle class and that everyone I knew was too. The only time we left east San Diego was to go to Disney Land in Anaheim or Amarillo, Texas, where my mother had a girlhood friend. Every summer we'd make the long, hot drive to Amarillo down route 66. All along route 66 there were billboards advertising an imminent turn-out for 'The Thing.' We never stopped. It was understood that 'The Thing' did not exist or, if it did, it wasn't worth seeing. It never occurred to me to suggest going anywhere else. I wonder

if I knew that people could share their thoughts. I suppose I learned that from books at some later point.

From the start, I was bookish. My mother said that on Christmas morning I would discard the dolls aside as quickly as possible, looking for a book. My mother read to me when I was small. I realize only now that I never saw her reading on her own. My father, when he was around, would sit in the chair he claimed for himself, the one we were not allowed to sit in, reading and drinking beer after beer. He never read to me. He might have enjoyed parenting a boy but he seemed to have no use for a daughter. By the time I was eleven or twelve, I understood on some level, that he had set up a petty patriarchy, a house where he had all the power he lacked in the larger world. I didn't like him, but I felt sorry for him. His little realm was, of course, part of a larger world that also demeaned women. The women I saw portrayed in popular entertainment in the 50s and 60s were highly sexualized, powerless, silly. If I had lived somewhere different, I might have been aware that accomplished, independent women existed, had always existed, despite the barriers they faced. But from Allied Gardens it wasn't easy to see a way out. As a young teenager, I don't think I even knew about 'beatniks,' but I had started wearing a lot of black. I had what we now call 'attitude.' If 'Goth' culture had existed then, I guess I would have been Goth. I aspired to be a juvenile delinquent or maybe an outlaw, partly because I was genuinely angry and partly because those identities were made. I did a few things for which I could have been sent to 'juvie,' as we called it then. I pictured myself as a bad boy because girls were boring.

I knew I was expected either to marry early or prepare for a career as a secretary, nurse, or school teacher. None of those trajectories appealed to my teen-age self. I had been writing poems since I was in first grade at least, but I didn't know there were living poets in the world. If there were, they were nowhere near me. I showed some of my (admittedly bad) poems to one of my high school English teachers and he responded, 'Women can't write poetry.' So how did I go from that almost total lack of encouragement to becoming a poet and university professor? I don't know how to account for it truthfully. I could say I was crazy stubborn. But I know luck was a big part of it too—that and good timing.

By the time I was in my late teens, change was in the air. I learned about it by reading magazines in my history class. The teacher always had a good stock of them; he wanted to keep us entertained. I remember reading a review of Betty Friedan's *The Feminine Mystique* in one. I went to the relatively well stocked public library near the candy store my mother ran (I had started working there during holidays). When I carried Friedan's book to the check-out desk, the librarian said, 'Oh, honey, you don't want that.'

But I did. And after that I read *The Second Sex*. What I really needed were girls or women with whom I could discuss what I was reading. I couldn't think of anyone. I did have one female friend, but she didn't share my taste in books. Still, I remember how one day we trailed a neighborhood woman around because we heard she had gone to college and we thought she might have something interesting to say. She didn't.

By the time my junior year came, I wasn't planning to go. I know that sounds crazy. I guess studying for one of the careers open to women didn't appeal to me. I was toying with the idea of running away somewhere more exciting so I didn't pay much attention when the SAT came and went. I don't recall being urged to take it. I don't recall seeing the guidance counselor at all. I don't even know if there was one at my school. Finally, though, I did have a change of heart. 'Lighting out for the territories' began to seem scary. At the last minute, I took the ACT instead—and also the National Merit Test. I did well enough on that to be given a medal in the school auditorium but not well enough to be given a scholarship. I applied to San Diego State College, part of the Cal State system, just a couple of miles from my parents' house. I don't recall consulting my parents about any of this. But they did pay the nominal fees and buy my books. There was no tuition then at California's public colleges. I had a little bit of money saved from my part-time work, but not enough to live on long. I started college commuting from Allied Gardens. Within six months, I was living with a fellow student named Chuck who, some years and a couple of separations later, would become my husband. His background was even less stable and poorer than mine. I kind of liked that. He'd earned money for school working as a dishwasher and bus boy. I know my parents quarreled when I moved in with him. My father wanted to cut me off. To him there were only 'good women' and 'bad women' and I had joined the second group. I had, to use his term, 'shacked up.' My mother continued to pay for my books and fees secretly. She must have been conflicted because she was an evangelical bible-belt Methodist. And she believed a good woman obeys her husband. But she was a loving mother too. To earn money, I got a part-time job as a 'teacher's aide' in a primarily African-American middle school. These kids had a lively skepticism about the advent of another white person who knew little about their culture—and they weren't afraid to show it. The teacher I was there to assist seemed totally burned out. I sat in the classroom grading quizzes while he criticized my proto-hippie clothes.

Still, college and my quasi-independence were heady. I took a poetry class and discovered that living poets *did* exist. I started using what money I had to buy anthologies. I had already found Emily Dickinson, now I discovered Denise Levertov, Robert Duncan, Robert Creeley. After two years at State (and a quarrel with Chuck), I decided to apply to Berkeley. I'd gotten good grades and there was a generous transfer program between California's State College system and UC, though I didn't know that then. I only knew that I wanted, somehow needed, to be in the Bay Area. I didn't expect to be accepted and, when I was, I didn't know how I would pay for it. I had enough money to get myself there, but then what? In the end, my parents decided to pay. I think I know why. They must have figured it was a good way to separate me from Chuck and our disreputable life. (This worked only in the short term). Like San Diego State, Berkeley charged no tuition when I arrived there in early 1969. Fees amounted to around $300 a year. For my parents, this wasn't 'nothing.' For them to pay that on top of books and room and board was something indeed. But it was possible. A real tuition would not have been.

When I arrived at Berkeley, or 'Cal,' as the natives sometimes call it, I found that Denise Levertov was a visiting lecturer there. What's more, she was looking for just one more student to fill her poetry writing course. She was holding interviews for this position. She must have been very popular indeed. I showed her a poem and told her I had moved to Berkeley to take her class. This was a desperate ploy. Actually, her presence was happy news to me. She opened her office door and told the line of waiting aspirants that she had 'found her student.' Once again, I was lucky. In that class, I befriended Ron Silliman's then wife Rochelle Nameroff and a grad student named David Melnick who was auditing. It turned out that the more radical cognoscenti (such as Ron Silliman and Barrett Watten) were studying with Bob Grenier. (He was lecturing there that year too.) But I would meet them soon enough. I had arrived at the right place at the right time. From there the rest of my life became conceivable—though still full of challenges.

By 1972, Chuck and I were living in San Francisco and I was going to the poetry readings that sprang up everywhere. I reconnected with Ron, by then divorced, and through him with the Bay Area poetry community. I was a grad student in creative writing at SF State and Chuck had a job in the receiving department of a Dalton's bookstore. It was still possible then to get by on one meager income. Just barely. Sometimes I had a part-time job; sometimes not. We always had a roommate. We ran very low on money by the end of the month. For a while we were on food stamps. I had friends who lived in more or less the same way, but most of them had other options. If I did, I couldn't see them. It turned out there was no job market for SF State grads who had yet to publish a book. For a while, I worked part time as a library assistant at the main library in SF. When they tried to make my job full-time, I quit. I thought I couldn't work full-time and also write. I still suspect that was accurate, but Ron probably scoffed. He was working full time and writing his fourth book, *Ketjak*. Of course, I was only able to make this choice because of Chuck's patience and generosity.

To make a long story very short, Chuck and I went back to San Diego when I got pregnant and our precarious life seemed unsustainable. He got a decent job as a buyer for the bookstore at San Diego State—where we first met. I began to get adjunct jobs at UCSD. We raised our son and I published books. Twenty years went by. At a conference, I met Suzanna Tamminen of Wesleyan who offered to publish my selected poems, *Veil*. Eileen Myles was hired at UCSD where she advocated (if not arranged) for me to get the tenured job which had just then opened up. Once more, I was lucky. At crucial times, people helped me. And, in my youth, the post-war economic safety net helped me too. In retrospect, I can see all the things that could have gone wrong, all the mostly metaphorical bullets I barely dodged. I'm pretty sure that, if I had come of age in similar circumstances in the 21^{st}century, in this time when people work two and three jobs to 'make ends meet' and even military families often qualify for food stamps, my story (it feels weird to call a life a story) would not have been possible.

In a minor contrast to Armantrout, Lyn Hejinian was born in the Bay Area, raised by a stable, academic family, and later attended Radcliffe. Hejinian titled her resource biography 'Profit and Precarity' in the language of progressive politics of our day. She professes in her accompanying note how 'uncomfortable' she feels writing about herself. Among the snags, her statement catches include the associations of her cisgender identity. 'Comfortable in skin?' Clearly, she is not so. In addition, her most widely read book is the aforementioned *My Life*, so as an individual she often colors outside the lines of progressive generalizations into more radical perspectives on society. While this fact might lead many to excessive reliance on individual uniqueness, no singular identity has supported the complications of her interactions with other individuals and surroundings.

Hejinian's resource autobiography produced a very different result than the other three. Her resource autobiography is sociocentric. I can also imagine an autobiography that only speaks of the individual psyche and personhood isolated from the world, but no one like that was selected.

Importantly, without mentioning poetry really, Hejinian speaks of herself as a resource adding a feedback connection to herself in an isomorphic notion of resources. Strangely, for me, she uses the term 'unnatural' in a moral context which I propose as a concept that assumes humans are not part of nature, a holdover from Platonic views of nature separated from humanity, a view only possible when seen through the organismic binary by itself as implied by her avoidance of most terms pointing to her habitat.

An aphorism in Book Four of Friedrich Nietzsche's *Daybreak* reads as follows:

> *Knowing one's circumstances.* —We can estimate our powers but not our *power*. Our circumstances do not only conceal and reveal it to us—no! they magnify and diminish it. One should regard oneself as a variable quantity whose capacity for achievement can under favorable circumstances perhaps equal the highest ever known: one should thus reflect on one's circumstances and spare no effort in observing them. (*Aphorism* 326; R.J. Hollingdale trans.)
>
> *One's circumstances*: there are many, some received (wanted or not), others achieved (advisedly or not), all interconnected but each producing different consequences in different contexts. Not all circumstances can be identified as resources, but the two are so intimately interconnected that, in my brief comments here, I won't try to disconnect them. Circumstances aren't necessarily dependent on a single resource, but much about them is determined by what resources (in the plural) are available at any given time and place. Likewise, what we call *resourcefulness* is to a large degree facilitated by available resources, either internal (physical and/or psychological 'strength' and/or 'weakness') or external (familial or more largely social and socio-political). Meanwhile, some resources can serve as a resource in one situation (under one set of circumstances) but prove detrimental in another.
>
> Somewhat arbitrarily, though I hope not illogically, I'd categorize the spheres to which my own particular resources belong as 1) the bio-physiological (bodily), 2) the private (mental), and 3) the social. That last is too large a category to be useful without first undergoing some subdivision—into, say, the familiative and affiliative (family and close friends), the professional (in my case the wider circle of intellectual and creative presences along with concomitant obligations, commitments, and ambitions), the financial (salary and savings as well as financial obligations and debt), and the political (abstract structures and systems that have real life effects

and variously generate promises and hopes as well as, more often these days, pernicious worries, anxieties, angers, etc.).

But I myself, too, am a resource, though of what kind and for what or whom depends now, as it always has, on time, place, and occasion.

I write this portrait-plus-landscape of a self, (more or less mine) in an historical moment predicted to have a dire future, as a married middle-class university employed white cis woman writer and academic who is 22 months short of being an octogenarian. Aging brings with it obvious resource problems. Typically, as one gets old one physically needs more and can give less; the resources available from one's own body diminish. And often this is true of one's own mental resources—one's ability to think and one's access to or control over one's emotions. My own bodily resources are all still more than sufficient for my daily needs and for most special occasions (though I stay away from protest actions in which being able to run well and fast could be important) and I'm still able to think and still willing to feel. And I do that thinking and feeling from within a strong family and an immediate and multigenerational circle of friends with whom intensity of affiliation has been and remains active and exciting.

But even as that thinking and feeling, along with the activities that accompany them, are pleasing to me and seem to have creative force and remain productive, so that my value as a resource of some sort continues, at least locally, nonetheless there is no escaping the fact that we are all living in destructive circumstances. And both these circumstances and the knowledge that they are largely the product of human cruelty, greed, and stupidity have profoundly debilitating effects. I see this in the lives of friends, students, family members, colleagues, and community members, all of them living in what they feel to be a state of precarity, whether financial or meteorological or political or social or temporal. All of us are living in circumstances that are not only unnatural but contra naturam. My own sense of what's historically true—of what is or was or could be real—whether of the present, the future, or the past, is now increasingly informed by disillusionment and disappointment. And attrition of confidence has been perhaps more important even than attrition of hope, wonder, and gratitude.

Of course, all four of those (confidence, hope, wonder, and gratitude) are resources far more likely to be available to the privileged (the lucky) than to those whose circumstances from the start are stressed, damaged, and constrained. But I wonder, just how beneficial a power is luck, finally? Does it make or mar circumstances, augment or deplete resources?

In an interview conducted by literary critic Thomas Fink with me, he said:

> For me, 'autobiographical' collage-effects of Hejinian's *My Life* and its sequels undermine individualism, providing evidence of how the 'organismic binary' falsely separates 'inside' (self) from 'outside' (world). In these texts, the social realm and the environment are often posited as constituting her experience. Hejinian's tone is often too anxious for feasting implied by your term 'cornucopia.' She constantly doubts the validity of her perceptions/language and indicates that her skill is never enough to control the poem. She does not evince confidence in her inquiries' long-term successes. For Hejinian, the process of interpretation of herself and/in the world is never-ending—always subject to failure and limitation.[1]

Differences and similarities of interpretations! Fink thinks doubt means austerity. I suggest instead that doubt fosters accumulation as any extended piece of writing attests. Hejinian separates herself for other writers within the language group by her autobiographical form. Within that form, she makes it clear that agency is a problem in that the individual is built in fragments and cannot be embraced as a whole without also accepting other problematic constructions. She differentiates her way of writing about herself from the usual memoirist who consider the self as a unit, that is, outside the language group.

Instead of the unitary self, *My Life* changes as the writer changes. Though not true of all of Hejinian's work, this particular work moves her toward an alliance with egalitarian writers in the matrix. Her identity changes with the role the writer plays in each text. Her effort to use the continually expanding text of *My Life* to, as she says, 'relinquish control,' seems only partially true. Expanding her text as events occur asserts her control over it, carrying the autobiography with her, reigniting her identity as a writer in control of the resources of writing (techniques among them) and the connections between the events of her life. Her agency is not relinquished in the way it would have been if she simply wrote a book and left it to the readership to read it differently over time. In some ways, then, *My Life*'s method increases her control over the text; that's what I mean by skill. I would not mistake her modesty for repudiation of her skill, but rather modesty as a pose she takes in this text. I'd also suggest that her focus on social being is central to her poetics.

Hejinian's resource autobiography focuses on herself in a dual role as an individual and social being. She doesn't write about her habitat or herself in an ecosystem as if that condition were fully understood or simply not worth discussion. As an individual organism and language writer, she twists terminology to include environmental vocabulary like 'resource' in her personal identity and within the social model she articulates. By doing so, she makes clear that her idea of self operates within Humanism: the person as an individual and within society. She does not connect identity to her surroundings, an odd omission considering the physical beauty of the Bay Area, the area's high cultural level of interaction with the environment, and the association of her politics with those protesting wild fires consuming the coast. Can lifestyles of individuals survive continued separation of who we are and how we live from our theories of language and poetry?

I suggest that individual identity conflicts less with its surroundings by understanding one's self as an organism, working with a diverse set of identifying roles, as a social being integrated into multiple groups, as an entity within several ecosystems, and as a connective link to various characteristics of other individuals, groups, and ecosystems of the planet and of ideas. Individual identities also operate through writing styles. These writing tendencies live side by side but people continue to defend them as organisms. One or two groups of writers might seem more critical to you at this time, but no one wants to destroy the water and atmosphere that support writing. Consider for a moment the kind of world that still supports life with human population dominance and human population dependencies. Slowing climate change depends on acting in concert.

Acting together assumes that there is a framework with flexible connection points where fusion of fragments can occur. The environmental model contends that the default condition of existence is multiple with different points of view about habitats, politics, self-interests, and languages. Our similarities—those isomorphic relations—don't prevent our focusing on difference to validate the organismic binary and its psychic and social parallels. Only occasionally do these varieties of roles come together as a single solution.

These four women who are related by current class affiliation, approximate age, and a high level of professional expertise have widely different points of view. Purely social assumptions about common class assumptions are submerged except for Hejinian. The configurations of organism and surroundings, as well as the complexity of experience, make a social identity unproductive by itself in determining one's characteristics.

Three of the four of these women accepted my simple instructions quite literally writing narratives of their lives generally based around their use of resources and their perceived availability. Hejinian turned the request for an autobiography of the availability of resources on its head by only addressing personal and social resources while ignoring her surroundings. As a result, she is forced to twist terms to avoid the obvious connections to her surroundings. While this was contrary to my expectation, it allowed her to use the term resources in unusual and instructive ways.

Tawasil has a similar outlook to van Norren regarding her responsibilities to address questions of resource management professionally in writing although through different disciplines. We can see in their pieces a similar way of speaking about the self even though neither are native English speakers. The details in these autobiographies are more specific than the categories in the matrix of the views of nature. Treating the autobiographies by themselves without reference to her other work enhances the competition of ideas visible in Hejinian's mode of address. As a result, she speaks directly about the conflicted behavior and conflicts in thinking that we all have to deal with while Armantrout raises fewer questions about the context of her narrative.

The plethora of details of self-examination constrain certainty for individuals, although Hejinian wishes that weren't the case. From the perspective of information theory, Erkki Patokorpi points out that "more knowledge increases uncertainty because it expands the sample, which in turn increases complexity."[2] But cultural groups defend opinions like organisms defend their bodies. I don't think I need further proof of this point to you, connected reader.

Living creatures will always try to avoid eliminating the organismic binary from culture; it's essential to having a cohesive body densely connected as an organism. We must address certain threats with one component of our identity—Black Lives Matter and police security, class hierarchy and social mobility, freedom and equality, and elite privilege (oligarchy, dominant race, corporate hierarchy, individual taste) and a balanced society. You know them well.

From here, I want to reveal certain options for resources supporting identity that are not part of our daily discourse. According to Afsaneh Najmabadi, the Iranian government, for example, subsidizes sexual reassignment surgery to normalize genders, to suppress marginal and cross-border identities. Their society seeks clear definitions in a world that may not be as distinct as Koranic categories.[3] In North America and Europe, reassignments are purchased and consistent with those market economies.

Combinations dominate all discourses. Physically, chimeras are not only likely (they are not the pathetic Halflings of myth), but rather fluid personae based on role. As syntax, they form vital combinations like America where more than 300 languages are spoken today. Modernism smashes cultures together in artists like Pablo Picasso and Sonia Delaunay in the last section of "La Prose du Transsibérien et de la Petite Jehanne de France" 1913. Combinations abound from lichen to humans: 90% of our cells are non-human and much of that non-human identity resides in your gut. So, think again when you talk about gut reactions. Who is doing the thinking?

In poetry, chimera as pointed out in the chapter on metaphor is commonplace as in Armantrout's skiing string instruments: "violins pursue / the downhill course."[4] Or her received syntactical discord, "Wild vista / inside or near the home."[5] Or surrealist juxtaposition or play with automatic writing that forms unusual, but since they are automatic, hardly unnatural, conditional combinations. Or Stein's aforementioned cow fault.

In society, chimera is commonplace. Many people play multiple roles from spies to laborers forced to work multiple jobs to pay the rent to anyone realizing that they think differently in different roles at different times of the day. Society as an organism in Aristotle, Marsilius of Padua, Fredrich Hegel, and Andrija Štampar is chimerical and well defended, not an illusion but a combination. We have to reconcile our differences to cross the street and change society to support the biosphere as the biosphere has supported society. As society changes, humanist identities do not disappear (we still emphasize and think of some parts of ourselves when we take any specific role) but they are circumscribed within the frame of environmental identity.

Social Syntax II: Linkages & Connectors

This chapter details shared characteristics of individual and group constructions. The network model of environment specifies connections between individuals, groups, and surroundings. Organisms and groups occasionally project their characteristics onto each other. These assumed connections, built between groups like classes, genders, races, businesses, governments, and ideologies, shape different views and characterizations of a group's surroundings. Connections between individuals and groups influence how they view the biosphere. A charity supporting poor people focuses donors on issues of ethical resource management. Labor unions identify limited resources to protect their members from competition while promoting their members' energy and skill as virtually boundless. Poets are both Shelley's "unacknowledged legislators" and Dickinson's "Not to discover weakness is / The Artifice of strength—."

* * *

Many organisms and ideas tend to gather and then function as a group. From microbial mats to beehives to human cities to poetic affiliations and even ideas about poetry, political economy, and environmentalism, groups and clusters construct forms similar to what we call organisms, similar interactions at different scales. As if they were individuals, operational groups distinguish themselves, create boundaries, interact with other groups, and evolve as a group for security and to reinforce their component organisms.

J. Sherry, *Selfie*, https://doi.org/10.1007/978-981-19-4870-1_17

Extraneous thoughts
pop into my mind
like quirky characters
of poets in groups.

While the connections within and between groups tend to be weaker and less trafficked than the connections that define biological organisms, some loosely structured groups last far longer than the organisms that constitute them. Ant colonies survive several generations of queens. Nation states have created connections including bureaucracies, constitutions, and trade routes that last hundreds of years. The US and the European Union perform as two collectives with different strengths of connection between their parts. Cultural entities strengthened by their trade routes and habitat occasionally last thousands of years like Pharaonic Egypt and the Nile and China with its rivers and the Silk Road. European capitalism has so far developed as an extra-national unit for 500 years and appears now to unite the globe albeit with abstentions and alternatives (Figs. 1 and 2).

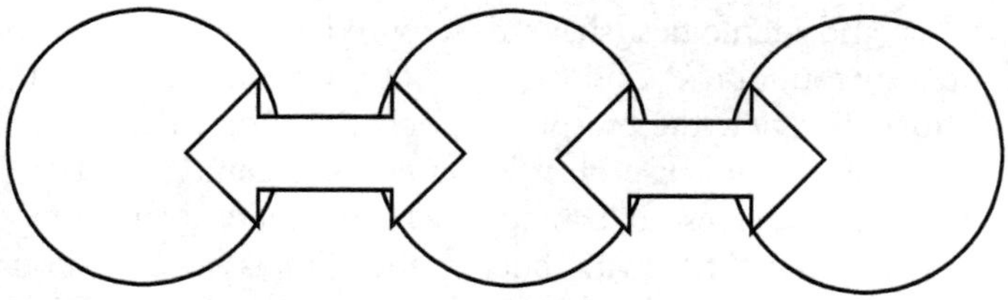

Fig. 1 A single organism with traffic across strong, bidirectional connectors

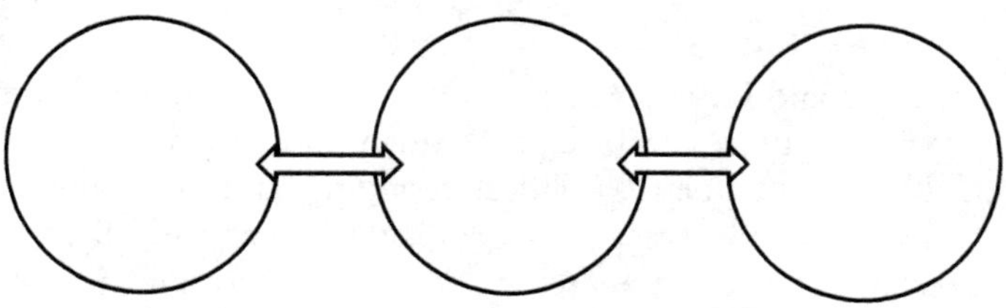

Fig. 2 A national or cultural group with traffic across weak, but still bidirectional connections

In general, groups that migrate strong connections within the group. Mennonites have survived as a tight-knit group, since the seventeenth century. Isolation entices and soothes especially for those with Thompson's autonomous and fatalist views of nature. During some periods, ghettoized groups infiltrate the larger society and, while pursuing insular practices, have a wide influence like Armenians and Jews with stronger and more heavily trafficked connectors within groups than between groups. Many national groups have been assimilated into the US such as Irish and German immigrants while retaining cultural links such as holidays, slang, and food. Some US groups have not been assimilated as well such as Black and Brown communities who are often isolated in ghettos with a denser set of connectors within the group than those to the larger society.

Creating opportunities to participate in the general society has resulted in increased wealth for individuals who share characteristics with the larger society. In China, Han groups thrive at the expense of non-Han peoples. In India, hundreds of Hindu groups mingle while Muslims are separated and threatened. The connecting links are many and varied and much type has been printed on the whys and wherefores of peoples' success and suffering based on their connections to dominant groups. Such empirical generalizations are useful, but are also widely misused to characterize individuals and reduce their access to resources.

Where does this busy, connective collectivity stop? It is limited by the number and throughput of the physical connections operating in a given society. For example, the notion of the group operates from the family to the clan to larger groups of similar and similarly inclined individuals like corporations, nations, and affinity groups like tendencies in poetry. The US had substantial traffic across the connections between classes in its period of expansion and after World War II, but now social mobility has decreased.[1]

Mobility between classes decreases as the linkages change their characteristics. European class connections were less traveled in the past but are now opening up as social mobility increases in the EU compared to the US. Nations emerging from colonialism like China and India are developing a middle class through employment channels, increased urbanization, and government regulation, wider education, and global trade. At the same time, globalization also builds a tier of ultra-rich families that undermines the rise of the middle class.

Income inequality, as Piketty points out in *Capital in the 21st Century*, remains one of the primary inhibitors of economic growth.[2] The rise of multiple classes and political points of view with access to communications resources presages conflict. The Covid-19 plague strengthened biological and chosen family networks and weakened nations. It strengthened networks within nations compared to connections between them. These changes to the level of equality and lines of communication between individuals and groups bidirectionally influence legislation, cultural boundaries between rich and poor, and technology change. Reframing these ideas as networks promoted the environmental model when events supported the network metaphor.

In the poetry world, there are various connection types with different characteristics. Poetry networks, understood as group identities, improve poets' relationships to their immediate coterie, habitats, and the biosphere as a whole. Connections used in this specific multifocal way have not been widely discussed in literary theory but rather mainly addressed in the abstractions of communications theory and anthropology. As presented in the chapter "Networks of Metaphor," connections operate in the way metaphors work in poetry by linking things and concepts in novel and unexpected ways through scalar similarity rather than the one-dimensional boundaries of intellectual and real property.

The close personal connections between Beat poets contrast with language writers' communications through text exchange and letters. The Bay Area language writers had closer personal relations, many having grown up together, than the NY language writers. These relationships are well documented.[3] But the development of language-centered writing's political concepts occurred in letters between Andrews, Bernstein, and Silliman.[4] First, a habitat linked the Bay Area writers' friendships and then epistolary connections built the ideological framework with the East Coast.

As a more general case, organisms and groups share resources for reproduction, productivity, social activity, aesthetics, and survival. To manage the division of these resources, especially surpluses and deficits, leadership maneuvers characteristics that citizens share toward sustainable societies. Documents like *The Federalist Papers* or *I Ching* show political and cultural networks in the making. The resulting societies all develop ways of handling surplus whether it occurs from increasing the overall stock of resources, exploiting labor, new technology, or a combination. Both labor and technology as sources of surplus represent risks to ecosystems.

The current increase in poetry production in the US, for example, occurs through several channels. First, increase results from an influx of money from MFA students. These students flock to universities, paying high fees with no more than a 15% prospect of a job in that field. They support the cultural myth of poetic value through their affiliations with poetry reading, poetry peer groups, and poetry education. The ability to isolate poetry from the larger instrumentalities of their education attracts young people alienated from material careers. Also, an increase in available themes and methods for writing poetry was introduced by the many postwar poetry groups, each with its own tactics and forms.

Innovations of form and new institutions attract young people. Black Mountain had its college. NY School writers built institutions like the Poetry Project and the Kerouac School to connect writers and support leading poets. Through their arty world ethos, these institutions use the low-cost labor of poets to broaden the reach of the NY School's network. Language writers began using free labor and state and federal grant money. Later language writers affiliated themselves with academic institutions for income, labor, and institutional support, thus building networks of low-paid academics to disseminate its ideas. Language writing also built digital archives of experimental poetry at Buffalo's Electronic Poetry Center and PennSound. My own small institution, Segue Foundation, is supported by exploiting personal connections, sales of books through networks like Small Press Distribution, government grants, the sale of its archive to the Columbia University Library, and the low-cost labor of poets.

Taste and evaluation in poetry are managed by group affiliations, publications, readings, and concepts developed within the group and through external theory. All are promoted to the group through available networks both point-to-point and broadcast. Having a strong academic presence with its teacher network helps distribute ideas of taste and evaluation.

Poets retain a strong sense of self in spite of group affiliation. Both autonomous individual poets, who view their surroundings as controlled by skill, and ineffectual fatalist poets often build cultures of resistance to oligarchy and standards. On the other hand, oligarchs of all political persuasions seek to manage large populations through a combination strengthening group affiliations to states and corporations while isolating individuals to suppress creation of interim institutions that reflect local interests.

This resistance sometimes creates change by divergence when a poet writes something against a prior poetry as in Grenier's statement against the NY School, "I hate speech."[5] Grenier's rhetoric tapped into resistance that Bay Area poets felt to the geographic literary mainstream in New York and diverged with a contrasting statement allowing a renewed connection to be established on the page as opposed to connection through voice. English Romantic poetry remains an obvious case of resistance, but its link to individual genius as its motive force supported both colonial and capitalist adventurism. Language poetry has merged over time through the academy into the NY School and vice versa as Ashbery has been referred to more than once as a language poet.[6] Categories of poetry come and go like species, but ideas are more virulent than species which is why I think poetry can help with climate change.

* * *

The environmental model assumes we all connect to, love, and support our biosphere because it supports us, but it is not a priority or frequently traveled connection for many people, especially those whose free and unfettered access to resources provides their income. Such individuals and groups often take resources as property and disregard peripheral effects like climate change. On the other hand, many poets want to talk only about social and personal situations, citing "unnatural" human behavior even though the phrase is a non sequitur. We saw that word, unnatural, used in Hejinian's resource autobiography when she wants to emphasize an ethical point. Many others with moral and social assertions in their poetry, living mostly indoors, disconnected from surroundings other than books, computers, and domestic and career anxieties, continue to ignore the biosphere, blocking the similarities of connection between any individual and its habitat. This is so across class lines and race lines.

Obstructing links to habitat and creating gates with constraints can readily be associated with capital accumulation because the primary contributors to climate change are not individuals but groups of corporations in the six polluting industries. Yet these corporations are supported by loggers, assembly line workers, miners, agricultural workers, and many others whose livelihood depends on direct access to resources. These corporate leaders largely ignore workers except to treat them like resources as in the term Human Resources.

Little by little even these individuals and corporations are getting the message that sustainable income depends on favorable habitat. Is there an important reason to find common ground in environmentalism? There certainly is, but it's unlikely that opinions about options in different cultures will be same. People have developed multiple solutions to unfriendly ecosystems that climate change might create. Windy places use wind mills, sunny places use solar panels, and volcanic lands use water and heat to generate power. Animal husbandry thrives in places unsuited to growing vegetables. Culture promotes capital intensive agriculture in wealthy countries and methods like the system of rice intensification in poorer countries. While conflict and competition have been highlighted as the forces of change by colonial theorists like Darwin, mutual aid has been emphasized by idealists like Kropotkin. Poets thrive on speech in social communities and written language in bookish and financial cultures. The success of solutions in both providing for people and not destroying the ecology in the process is governed through network connections between those solutions and climate.

When conditions are unsuitable to the well-being of certain roles of individuals, those jobs die out like blacksmiths, relocate like Armenians after genocide, and build sub-cultures like Black slaves and later different people of color in the US that work to improve living conditions and assert identities that comfort them in their groups and as individuals. In either case, the mechanism of change is usually divergence in both society and biology as Darwin, Gould, and Wilson point out. The network is built by a combination of natural selection, sexual selection as a form of aesthetics, and genetics.

Nevertheless, some groups stop functioning for internal and external reasons. In the poetry world, a power couple divorces, breaking their link, and the coterie of Flarf poetry expires. A source of water dries up and a society recedes as we see in California in this decade. Extinction may be very much the question of the day. These factors operate between individuals and groups with a combination of mutual aid and competition resulting in divergences that may also be called innovation when viewed as a kind of productivity or creation when connected to biological change. Differences do not eliminate similarities of shape and function as in wings for mammals and birds, as in protectionism for organisms and nation states, as in antagonisms around the moral implications of style among writers, dogmatic subjectivity vs scientism.

Divergent sub-groups establish their own supportive cultures. Sometimes they thrive and build their own societies like the Normans in England. But a related group of Normans failed to establish a permanent footing in Sicily to a large extent because the channels between Sicily and Normandy were not as strong or easy to maintain as to England just across the channel. And England is easier to protect than Sicily which connects to many other nearby locations and was sought as a power base by several other peoples. Multiple sub-groups dominate poetry communities in postwar literary group hybrids that diverge with astonishing rapidity and assertion of minor differences.

Climate change has not sufficiently affected US, Indian, and Chinese leadership's ability to control their populations for their oligarchies to address the risks and implement climate solutions with the urgency demanded by ecologists and climate scientists fully and publicly. Even though many projects have been started to arrest climate change, action is subdued and masked by public discussion of whether climate change is real, urgent, and anthropogenic. The gate for change resists opening.

How much do conditions need to change? How much flow of information from the biosphere directly and from cultural channels indirectly showing immanent societal collapse before the mainstream culture starts to promote environmental models to address the challenges to face and sacrifices required to arrest climate change? Is it desirable to have mainstream environmentalism before sustainable solutions are available and cost effective? And what happens to cultural diversity then? Can we have mainstream sustainable solutions while government is dominated by leadership dependent on the oil industry for funding? Yet even as I edit this chapter the *New York Times* proclaims anthropogenic climate change causes flooding of low-lying areas.[7] What happens to civil agreement in the US if the mainstream starts to ring the alarm of climate change? Already, general consent is compromised by changing power structures as climate and consequent social imbalances start to destabilize Western society. Conservatives, feeling their stability threatened, are already building institutions and movements to resist change.

As mentioned above, poets who established group connections sometimes break their affiliations with each other when one of them begins to become successful or even appears become successful. When connections are stressed by inequality, the expectations of the individuals on both sides of the approximately equals sign (≈) begin to change. Environmental views of connectivity might transform these like/don't like interactions by reminding successful workers and poets of their actual conditions and group affiliations. Can society insist that people owe the group that enabled their success? How can they be convinced? Who remembers where they came from and who wants to forget?

Poetics that subordinates the multiple components of identity, language, and consciousness to only one or two of them protects us from some exclusionary power structures with alternatives. A young, white, male experimental poet like Bernstein found more opportunities for publication within the community of experimental writers with their own publications like Roof and Sun & Moon than in the larger pool of all poetry that includes *Poetry* magazine and other mainstream publications. As the number of his connections developed through writing, speaking, organizing, institution building, and academic affiliation, his options in other literary groups increased.

As a young, Black, experimental, woman poet, Erica Hunt found publishing opportunities within the community of language writing due to her focus on alternative strategies and a strong political awareness she shared with the group. Roof published her first full length book, *Local History*. Her poetry was valued among experimentalists, but readership increased significantly when she got an academic job, affiliated increasingly with Black writing groups and experimental writers of color, and began to write more thematically in the context of race while retaining her roots among language writers. More connections and addressing more of her identity components increased her visibility.

The complexity of these connections looks a lot like the complexity of interaction in any ecosystem. From a stylistic perspective, even white male hegemony dissolves through entropic fragmentation in avant garde satires (McCaffery's translation of the opening of "The Communist Manifesto" into West Riding of Yorkshire dialect), successive formalisms (New Formalism followed by Bernstein's parodic "Nude Formalism" and others), serial binaries (attacks by any group on any other group), and potentials for well-defined sub-groups to support ecosystems with environmentalism both independently and collectively. The accuracy of the collective does not assure its success any more than my attention to it.

Each privileged and emerging group seeks the power of other groups with few exceptions in either an environmental or a capitalist model. Sub-Comandante Marcos may have been one of those exceptions. I don't see that issue being generated from the poets themselves or from their poetry, but from their professorial role as defenders of cultural faith and from the way cultural identity has been built by looking backward. I suggest that because of its low profile in the mainstream culture poetry has a better chance to initiate fundamental cultural change than movies and rock'n'roll partly because poetry has little contemporary credibility as knowledge and meaning increasingly adhere to images.

Identity's Constructive & Connective Ecosystem

This chapter presents seven detailed examples of connection to clarify how material joinery in an environmental model works to form the borderlands between and the apparently discrete entities of people, things, ideas, and components of language. The modeling process in the brain establishes replicas at different scales of the biosphere including the whole ring of life on the planet. Writers in all disciplines reuse that modeling process to describe our world. Each example tries to show how humanism's binaries can be replaced by a more detailed, multi-channel model.

* * *

Connecting processes compose me, but I have trouble accepting it because I'm continuously separating myself from my surroundings and other people even while I'm functioning with them. My body pushed me to see myself as one, and as I grew, other people became another. Gradually, I understood social units because I sensed so many similarities and my parents trained me to compete in sport, school, and social settings. To see my surroundings as similar to yours was not difficult, but at the same time I noticed that you saw differently than I did. Many years passed before I started to understand that these differentiations and attachments were connections of different types. I started Segue Foundation as a general idea of connection. *Selfie* specifies connections and my affiliation with L = A = G = U = A = G = E magazine and *ecopoetics journal* enhanced

J. Sherry, *Selfie*, https://doi.org/10.1007/978-981-19-4870-1_18

the idea of connection even though the equal signs represent a mathematically improbable environmental solution and the abbreviation eco misrepresents the longer, more congested traffic that we must live with to survive.

Reading my mind as it operates in discrete quanta of awareness about ongoing processes increased my difficulty in acknowledging the connection of my actions to my surroundings. I kept track of events through distinct episodes with a recognizable syntax. Apparently, episodic memory retains more when tied to architecture, mnemonics, acronyms, chunking, rhyme, and stories, that is, when a model is present to help order events. This is so especially for people for whom a change in how identity relates to ecosystems is most urgently required. It is true for humans and some other species.[1]

Thinking through the environmental model connects me to my surroundings as distinct from the humanist opposition of self/other, inside/outside, and subject/object that isolates me from others. Similar structures like story, song, and shape of poems connect social groups, but as poetry becomes more complex, different languages create different poetries. Thus, these connections change shape, method, and matters or themes. Here are some detailed examples of connection.

Example 1: Attention to and awareness of climate change through channels connecting different parts of the brain.

People complain how their minds are affected by the weather, but only recently have some begun to focus on the channels by which consciousness in turn affects the world. And of course, this is what I mean by ecology. Recent discussions by Graziano and others about the impact of sensory signals on brains identify the channels that build awareness. How does that work?

Living organisms have been able to select which signals to focus on for more than 500 million years. Nervous systems have had an internal model of attention for 350 million years. Animals have had a social attribution of awareness for six million years at least. Human awareness is established through connections between the area of the brain where sense data collects in order to make us pay attention and a model of attention in another area of the brain. Our awareness can then pay attention to both the external information and our thoughts. It is not a modern phenomenon but has been widely mystified.[2]

Graziano depicts consciousness and how it is achieved along the pathways between awareness and the information one is aware of.[3] In order to actually be aware, our brain has to connect the information being processed through neurons with the model of attention that is built into the brain. We may be sure that we have awareness as Descartes pointed out in *Discourse on Method*, but little attention has been paid to how "the feeling [of consciousness] itself impact[s] the neuronal machinery, such that the presence of awareness can be reported."[4] If awareness is an informational model built into the brain, some of the mystery about consciousness dissolves. Graziano draws a similar figure of consciousness, showing how science hasn't really discussed connection back to the senses[5] (Fig. 1)

> It is with [connector to senses] that many of the common notions of awareness fail. It is one thing to theorize about [connector to awareness], about how the functioning of the brain might result in awareness. But if your theory lacks [a connector back to the senses], if it fails to explain how the emergent awareness can physically cause specific signals in the specific neurons [connectors], such that speech can occur, then your theory fails to explain the one known objective property of awareness: we can at least sometimes say that we have it.[6]

How do these abstractions develop? Our minds make inferences that go far beyond the scarce information available. As discussed above, children only need to see one or two horses before they always recognize a horse as a horse. They develop models about things that they sense easily and

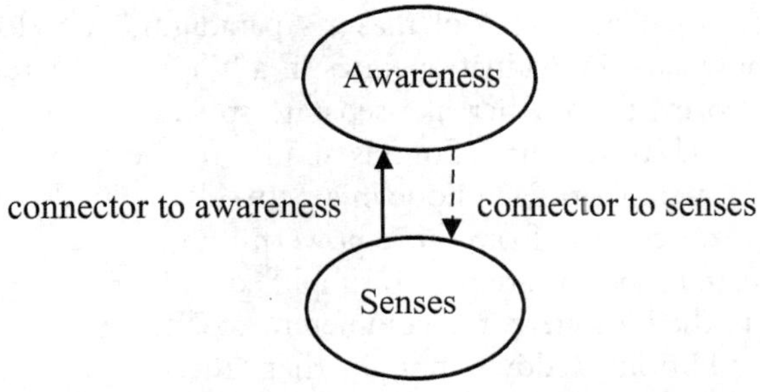

Fig. 1 Consciousness as awareness of the senses

their skill at inferring makes children quickly able to distinguish between a horse and a cow.[7]

Tenenbaum's abstract model of the horse allows us to generalize through Graziano's attention schema inside the brain. The model establishes a channel between our thoughts and what we see in the world. Through two different actions—cognitive processing and sensory identification of our surroundings—the biosphere, at this mental scale, constructs our thoughts using similar modeling methods. We build models both to be aware and to identify ourselves, our societies and our surroundings.

So, it makes sense that we would use modeling to construct our understanding of other complex aspects of our surroundings like models of climate, social structure, and writing. It's a skill we have developed through evolution. Modeling, a physical process, is built into the way biology works and is not only a scientific and logical simplification. Graziano and Tenenbaum's teams invert the journalistic expectation that human rationality models complex nature. They point out that the biosphere uses models. Similarly and unsurprisingly, so do our thought processes. These models, one of which is metaphor, are physical interactions between the brain and its surroundings.

These schemas, different in both Graziano and Tenenbaum, also show how Reddy's concern about the interchangeability of concepts that are connected—ideas, thoughts, meanings, and feelings in his examples—operate as shorthand for how the schemas work in detail. The verbs—put into, get across, give—shorten the rather complex processes of connection that was both not understood very well. These verbs also acted to avoid lengthy, inefficient, and repetitive descriptions of connection and transmission at the time these phrases started to be used in English.[8]

Reddy's alternative, "the tool maker's paradigm," develops another schema for how each individual operates in a "slightly different environment" and also exists together like separate spokes of a wheel with the hub of the wheel as schema.[9] Sounds similar to the web simplification just a different metaphor. Each node in a network might also act as a hub with more spokes coming from more powerful hubs.

Environmental schemas are not so simplified and by linking more fully with models in the biosphere, the completeness of metaphoric connection increases. In addition, Reddy's thought that "there is absolutely no way for the people to visit each other's environments, or even to exchange samples of the things they construct"[10] does not conform with what we know about the myriad of communications between individuals at many levels of language.

I might agree that some visits between individuals are "crude," but the expectation that exact and complete instances of transmission are required doesn't conform with the reasons we need to speak with, write to, and touch each other. In fact, too much information that includes all the perceptions, thoughts, and the schema confuses the issues needed to be communicated and exchanged.

Fortunately for poetry as a carrier of environmentalism (shorthand), writers built multiple schemas. Through its channels poetry long ago established what science is only beginning to acknowledge as a mental process. Blake connects the word and the world in his 1794 poem "London":

In every cry of every Man,
In every Infants cry of fear,
In every voice: in every ban,
The mind-forg'd manacles I hear

Blake hears the repressive social schema in the shorthand way people speak and cry, their bans and banns. Foucault picks up this pattern and feminist theory expands it. Materialist poets like Silliman echo these connections between the brain and the world by pointing out that the specific pen, the differences between speaking and writing, changes the poem: "8. This is not speech. I wrote it."[11] These kinds of connections link the materiality of language to the material world. Olson's "to build out of sound the wall / of a city"[12] turns the complaint of Blake's "London," the cries of residents, into a strategy for poetry to materialize the world.

Graziano materializes and expands on the idea of mental connectivity to differentiate and link awareness about the world to awareness about the self. If we say that the schema of attention in the brain links to information from the senses in another part of the brain to create consciousness, we can also say that the schema of attention links to information generated by mental processes to create self-awareness[13] (Fig. 2).

The brain uses the same attention schema to create the awareness of ourselves that it uses to create awareness about the world outside ourselves. If Graziano's model of awareness, even simplified, largely operates as he says, then theory of mind in non-humans seems likely especially in species operating with a high degree of social engagement.[14] This

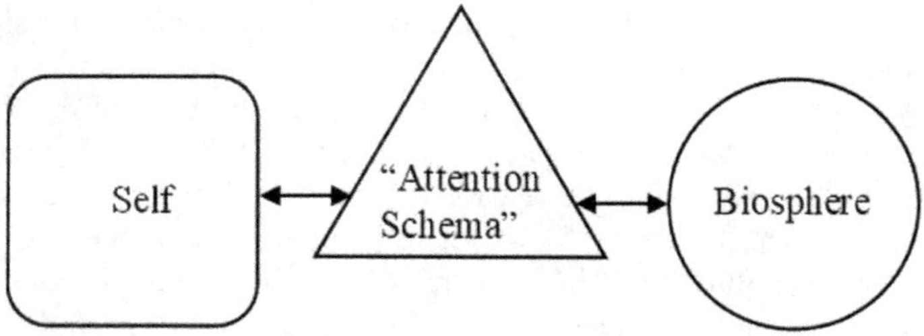

Fig. 2 Attention schema mediates self and surroundings

makes sense because Graziano shows that this attention schema is generated by the same parts of the brain that generate social awareness. Awareness of self, awareness of the world, and social engagement all light up the same parts of the brain.[15] Poetry exploits this attention schema by connecting awareness about the world, self-awareness, and social critique.

Example 2: Poet to Poet Connections.

Over time how people identify themselves and online life have converged with daily physical lives as isomorphism inexorably obliges the organismic binary to acknowledge its surroundings. The multiple entities comprising each human together make the internet an increasingly accurate reflection of ourselves. As people become more dependent on the internet, our real lives bend toward network personae. Fumbling on the computer is an act after all, and actions have consequences. The interactions in cyber- and other spaces help poets develop identities that are both language oriented and models of their aspirations.

The interactions of individual poets range beyond their style, nation, and behavior. In the poetry world, poets' relationships to families, to writing practice, and to language resources change frequently as new fashions of living appear in society and new stores of information became available from Bestiaries to Bibles, from compendia like the *I Ching*, *One Thousand and One Nights*, and *The Canon of Medicine* to Google search.

Until recently, however, poets' connections to audiences changed more slowly. Although the printing press vastly increased channels to audiences, the humanist connectors remained more or less linear from author to publisher to bookseller to reader. As society became more complex and interactive, these linear channels became less relevant to the manifold interactions of minds and societies. In recent years, however, many variations have occurred in poets' access to language resources, to databases that also operate at the intersections of network channels, and to audiences as well. Transport mechanisms—email, Facebook, Twitter, streaming media, websites, digital printing, and wide access to poetry education—travel through electronic and social networks that operate like biological and physical networks, like city streets, and carry information about poetry to expanded audiences. The number and character of connectors have expanded exponentially in the past forty years. Natural human inventions have overtaken other natural metaphors like the web.

In some ways, both writing style and the medium become inseparable, connected. McLuhan: "The content of any medium is always another medium."[16] This simplification, however, looks different refracted through this essay. Networks within networks certainly operate *all the way down*. Every turtle can be modeled in a limited way as a set of networks within networks. Any unifying one evolves to another network. Their connections collect to a thing that then disintegrates upon inspection and redefines singularity as a process. Any one isn't. Even the oneness of them evolves. And yet, in spite of my irritation at navel gazing when action is needed to avoid devolution, sitting and looking at a scaling model does produce specific and useful understanding of itself, group, surroundings, and their connections.

The environmental model reinforces not only a scalable method, but also the diverse identities, perceptions of identity, and functions of those identities at the intersections of networks that people see and use to survive. After all, the purpose of the environmental model is to encourage species survival at sustainable intersections. Some characteristics of how network channels cross have been termed "intersectionality" by Kimberlé Crenshaw.[17] Intersectionality includes the layers of identity built up by social discrimination. Along with the nodes at intersections of networks, intersectionality characterizes how those nodes are affected by the network. Crenshaw's intersectionality draws attention to conditions suppressed by not only white male power, but also by middle-class, white feminism; that is, how groups of well-connected individuals suppress contributions by unaligned individuals and marginalized groups. As articulated by bell hooks, the emergence of intersectionality and other practical feminisms establishes environmental linkages.

> Privileged feminists have largely been unable to speak to, with, and for diverse groups of women because they either do not understand fully the

> inter-relatedness of sex, race, and class oppression or refuse to take this inter-relatedness seriously. Feminist analyses of woman's lot tend to focus exclusively on gender and do not provide a solid foundation on which to construct feminist theory. They reflect the dominant tendency in Western patriarchal minds to mystify woman's reality by insisting that gender is the sole determinant of woman's fate.[18]

What happens at the intersections? First, communication between the node—people, genders, races, roles, jobs, nations, animals, plants, energies, ideas, databases, and models—and the material carried through the network (another scale of intersectional interaction) identifies the node to allow access across its borderlands. These interactions vary for types of nodes like a nerve plexus, an internet router, introducing yourself at the beginning of a conversation, and a book with its cover image and back cover identification. Both discrimination and differentiation among nodes occur in this first step. One notices race, gender, and other characteristics and then inquires about possible connection to the other person. Identifying the node, person, and text according to custom both validates the protective tactics of the organismic binary and compels us to understand the shared activity through channels and across intersections.

Second, one will try to separate what is carried across the network that applies to the node from what goes through the node in its occasional role as a network channel. Some poets claim that they are merely carriers of network traffic and not creating it. "Not I, not I, but the wind that blows through me!" as D.H. Lawrence says. (Arthur Schopenhauer's musician is a "telephone of the beyond" and "ventriloquist of God.") In Vatic poetry, higher powers speak through poets. Automatic writing is a modern version by appealing to subconscious powers of creation. This happens frequently. In fact, much of what is said channels another's messages.

Third, the poems, nutrients, and other material carried across networks differentiate those acts that discriminate where equality should operate from those acts that save us from homogenization. The details of any communication follow the scalable associations of environmentalism except where the organismic binary interferes with traffic. Sometimes interference is simply a matter of crowding the network with too many intentions and consequent collisions, a caution about blame. As mentioned above, over-determination risks failure of environmental action through risk homeostasis and is one of the most serious threats to acting on the climate scale.

The license given to writers of my generation by counterculture poetry and by academic freedom has resulted in some not understanding individual and group boundaries, the careful codes used to cross those borderlands that are important to individual sensibilities in threatened communities. Poets focused on race, gender, and ability appropriately object to poets who take advantage of the freedom to write without concomitant responsibility/constraints/comprehensibility and attention to others.

The property model of style among free writers through consumer choice, misreading the Enlightenment, and the heightened focus on individualisms spreads from fields as diverse as medicine, politics, consumer products, and science about individual consciousness. We are in the midst of a renaissance exalting the individual without informing those individuals about the similarities of structure in complex ecosystems that occur at different scales, that constrain certain individual action in support of overall freedom of action. Myopia puts human-friendly climate at risk.

Example 3: Connections between Marriage and Homes.

All humanity shelters from storms, but different ecosystems impose different limits to housing. Igloos, straw huts on stilts, concrete blockhouses, and wood frames all serve a parallel purpose but connect to the ecosystems of the dwellers differently. Connections between dwellers, communities, and ecosystems also identify with specific aesthetics and rank. Marriage in industrialized countries can historically connect between families for continuing blood lines with property including houses being exchanged and transmitted through marriage. Gifts may be given to brides and grooms by the families as common property. Ownership can be individual and in common while the couple's relationship as husband and wife is established in the eyes of god and the state.

In Madagascar, Zafimaniry houses are directly tied to marriage. The identity of the couple as husband and wife develops with the house that the husband builds for the wife. The couple's house is situated near the husband's parent's house, but lower down and south. The marriage is strengthened by the birth of the first child and by the parents naming the child together, but weakened because the wife returns to her parents' house to give birth. There the child is tied to the mother's family by rituals such as burying the placenta there and marking the child's forehead with soot from the hearth.[19]

Before the mother returns to her husband's house, both he and his family must court the wife again. Only with the birth of the third child is the marriage cemented by the mother's mother helping with the birth in the husband's house and the child is marked with soot from the husband's hearth. Maurice Bloch:

> Only when children are born into it are the initial stages of marriage or of house creation complete.... But the house is also changing materially as a result of a different but linked form of conjugal work. When first built, the house is highly permeable to the outside.... The flimsy permeability diminishes with time. The Zafimaniry say that the house will then be gradually acquiring 'bones.' This refers to the massive wood planks that will, little by little, replace the woven bamboo. Ultimately the house will look a little like a Canadian log cabin, except that the wood is vertical.[20]

As the house grows over time, the marriage is strengthened and defined by the house building process that includes the brother of the husband and the sons as they grow. Bloch thinks it is right to consider house construction as "the continuation of the process of marriage."

The Zafimaniry are famous for their wood working, since the deforestation by colonial invaders in the eighteenth century. All wooden surfaces including the houses are festooned with carvings, expressing complex connections between houses and marriage as well as community.[21] The native species of heart-wood used in house construction appear as a connection to the strength of the marriage. The vagaries of affections and health mean that the channel between any individual house and married couple is neither continuous nor flowing equally between the house and the couple.

Example 4: Similarity of Solutions to Climate Change.

Communication of facts, statistics, and predictions about climate futures flows in one direction through media channels from scientists and climate deniers toward public opinion. Then, public opinion flows toward corporate leaders who watch the news and polls. They can mandate change in industrial processes and information flows to scientists and climate denial groups. (Some of the climate deniers such as executives of Exxon and Koch Industries also can demand change, if they are convinced

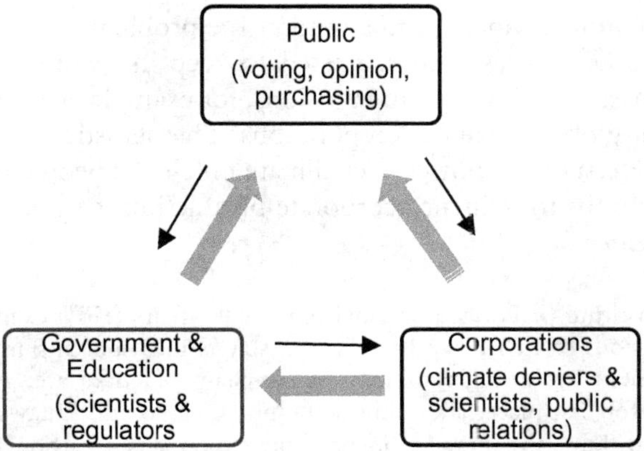

Fig. 3 Influence of public, corporations, and government

it will be profitable, by adjusting their supply chains and methods of resource extraction to reduce emissions.) The weakest flow extends from public opinion that lacks coordination toward corporate deniers in the six industries that are causing climate change (Fig. 3).

Public influences on government may not be as strong as news media suggest because the substantial, influential flows come from corporations to government through lobbying, media support, and donations to candidates. I have indicated a thin arrow of influence directly from the public to government because voters are driven largely by oligarchic, public/corporate, information sources; that is, few voters learn about the issues from independent sources. That said, bureaucracy does move inexorably toward effective climate policies. Flows influencing corporate polluters from both government and the public remain less strong as indicated by the thinner arrows. While corporate influence on government is strong, government has initiated few new controls on corporations in the past 40 years. Nearly all of the effective regulations on pollution have been agreed by the corporations themselves or as a result of oligarchic infighting.

Environmental movements have been less effective than corporate leaders considering that environmentalists have the entire biosphere supporting their efforts to reduce emissions. Global demonstrations of six million or more people at the September 2019 marches protested governmental and corporate inaction, but large segments of the American public remain unconvinced, India ignores the problem, and China, while initially making progress, has regressed to keep its economy humming. 70% of Chinese contributions to emissions, for example, are tied to manufacturing for global consumer corporations. This statistic should help the public to understand its power over climate unless the people build appropriate institutions to influence corporate production methods. Ma Jun in *Time Magazine*:

> .... the Institute of Public and Environmental Affairs (IPE) examined the climate actions of 118 IT and textile-industry brands sourcing from China to rank them by how green their supply-chain practices are. Apple and Nike tied for first place, and Chinese brands Lenovo and Huawei reached the top 30, but most brands did not take supply-chain carbon footprints into consideration. Barely any set supply-chain emission-reduction targets. Consequently, most of the top global brands may not be able to meet their climate commitments.[22]

Using public relations channels for changing public opinion has worked in the past on a large scale and it works in the present through funding by climate deniers. Bernays, Creel, and President Wilson changed the view of Europe toward the US after World War I. After World War II, global identity was promoted through the concepts like the "American Century" and "American Exceptionalism." In the late 1970s, American identity began to change again. Conservative groups that had been marginalized in the globalization process began to take control of local institutions and governments while financial and corporate leaders built a global network circumventing national borders. Increased trade with low wage countries increased the global labor pool leaving many communities without support but also benefitting many of the globe's poor nations at the expense of Europe and North America.

There is no reason to consider these changes out of reach for environmentalists except that the skill and resources to make those changes rest in the hands of the public relations firms and corporate media. The connections already in place are readily able to change leadership's opinions in the same way Bernays, Creel, and Wilson convinced leaders of Europe about democracy through what Bernays called "the engineering of consent." So far, these methods have been effectively used to prevent masses of people and leaders from addressing climate change. The flows through these channels can be altered with will and skill.

Scientists and the public can use effective public relations. Instead of confronting resistance by saying as the scientists say, *stop global warming now, it's bad for the environment*, Bernays' approach creates a climate of opinion that alters the ecosystem of opinion to build agreement and demand. Images of smoky skies and choking citizens can be juxtaposed to beautiful lawns and blue skies not filled with happy cigarette smokers as in the initial public relations campaigns, but children playing near farms filled with healthy crops, baby animals, solar panels, and wind turbines.

Fill the air with these images and their isomorphs at every level of culture in large numbers through the channels that are built to create public opinion toward corporate leaders. But convincing corporate leadership that their wealth, productive investments, and self-interest lie in a future where pollution is radically reduced requires another set of flows. Corporate leaders will not really be swayed by Bernays' happy approach to the public. Financial arguments may work better. How can this be accomplished?

Economic information shaping the opinions of wealthy individuals and corporate networks can be channeled to convince leaders that progressively lower profits will stem from continued production and consumption that pollute. These messages, constructed by public relations and economists, can be presented from accountants, consultants, and other finance leaders showing how profits will progressively decrease with business as usual.

These messages will be valued if they come to investors, business process owners, and political leadership through channels they already use and trust. Messages must show projections where pollution slows growth by suppressing demand and increasing costs of production beyond the ability to raise prices. Fossil fuel corporations then have the option to reduce their reliance on assets in the ground by purchasing solar, wind, hydro, and geothermal companies. Leaders of large corporations must be shown a transitional path from polluting to less-polluting production.

These messages must also come through technology to masses of mobile phone and computer users that reduces the need for movement of people and provides emotional, nutritional, and security support through reduced mobility. Electric engines replace internal combustion. These methods come as no surprise, but the solution begins with redirecting information flows through existing and new networks that convince leadership that arresting climate change is in their interest in the near future. Addressing vested interests remains the most difficult part of this equation because leadership, while already convinced, does not communicate its position to the larger population in order to retain control of institutions. What role does poetry play?

Example 5: Isomorphic Channels in Poetry Affinities.

Poetry connections and information transport mechanisms have changed more radically than either the relatively static view of audience or even frequent methodological change. We know from our daily lives that electronic media change the transport mechanisms of poetic works, but how do those mechanisms operate? What is their commonality and difference? Can computer networks be compared to human social networks, neural networks, and the carbon cycle? What are the problems with a communications model like environmentalism?

The poet erica kaufman shares certain components of her identity with me. We are both poets, politically progressive, US citizens, and born Jewish, but we don't share age, gender, sex, or profession. She and I connect strongly and bidirectionally through poetry as our primary affiliation exchanging many messages, written, spoken, and indirectly communicated through various poetry group affiliations such as the Poetry Project, Bard College, and through our publishing interests, she through Belladonna and Litmus Press and I through Roof and the Segue Foundation. We met through a poetry connection when she took over inventory and shipping from Africa Wayne at Roof Books in the building that Segue sponsored with poet Erica Hunt, filmmaker Abigail Child, and dancer Sally Silvers. Connections abound as phenomena assemble.

While we share many progressive political concerns, the national conversations are couched as secondary matters about freedom and privacy rather than a primary poetry connection which is more nuanced in its relationships to freedom and constraints as well as freedom and comprehensibility. Our connections around national affiliation are weak. (If we had met in the military or on a community board for the Lower East Side, political identity might be our primary connection.)

Jewish culture, while strongly related to how we look at language and reasoning, does not extend to a religious affiliation. Our history and culture as Jews are more important to us than our religious beliefs and Zionism which we agree establishes an illogical and unsupportable connection between religion and real estate. We are not afraid to separate history and culture from religion even though there's an important historical connection to belief. In fact, we had never spoken about religion until she read this essay to confirm how I have represented her.

The poetry channel on the other hand is robust, multidirectional, and direct. Prosodic tactics in our poetry, such as fragmented sentences, nonstandard syntax, and appropriation of popular culture and philosophy, also demonstrate this partite construction of ourselves and social interaction. We talk to each other as poets but use those secondary connections as well.[23]

The US administration under Donald Trump and other autocracies worldwide have encouraged erica kaufman and me to discuss our nationality more directly. I joined the American Poets Congress steering committee in 2017. In 2018, I began working for the campaign of Antonio Delgado in NY's 19th congressional district to increase the traffic through my long-term connection between poetry and politics. kaufman's job at Bard College allows her direct impact on poor students around the world to activate her political point of view. Our states of being are conditional: connections to any of those secondary channels could change if, for example, I suddenly expressed to her my views about America's reliance on gunboat diplomacy in south western Asia. kaufman's book *Instant Classic* does not contain the words America or Israel, but here is one reference from her introduction:

> Who owns language? Is language a body to liberate? Milton was undeniably not the most palatable of literary figures—publishing radical texts on both censorship and divorce in a time when that was not the genre of discourse one engaged in. In 2012, i am reluctant to name myself, to give in to the egotism that might be inferred from wanting to lie alongside John Milton. But, i am a woman, a lesbian, a Jew. In my daily routines, it is perhaps easier, within academia (or my version of academia) to not exist. i pass. i can pass. sometimes i choose to pass.[24]

When we talk about her identities that I don't share, the less traveled gender and sex connections, I tend to inquire about them, asking how she sees the world from those perspectives. Because of her good manners, she offers little without my inquiry.

This specific set of connections between erica kaufman and me emphasizes the complexity and diversity of reference between individuals, even group identity stimulated by social markers such as age, race, gender, and ability. This compound identity, operating as and through a network, might also describe an organism and a social group—similar (isomorphic) networks operate on different scales. Proposing these similar ideas at different scales is central to environmentalism in poetry. My interest in phrasing that we think about our surroundings like we think about ourselves may be a useful approach for poets who understand the similarities in poetry renditions of relations between self, society, and surroundings.

Example 6: Difficult-to-Travel Connections

Many indigenous origin myths use the generic word *people* or *the people* to inspire those narrating the myth and to exclude other groups to preserve resources. Environmentalists and the many other groups seeking a more inclusive hierarchy must find ways for all humans to become *the people* and subsequently for non-humans to be included in a more comprehensive classification to drive industrial policy. Building identities that can direct society toward inclusiveness travels complex routes through affinity groups from specific to general, a more difficult route than general to particular. The point about inclusion is evident but which specific activisms will be effective remains obscure. Here are three examples that have worked in rough terrain.

- Euro-American myths of productivity and progress drove nearly all of the world's cultures to adopt their model of productivity. Major populations including India and China participate in the West's techno-industrial ecosystem. More individuals increasingly had more free time until software absorbed the free time created by automation. The critique of the system has been largely that not everyone reaped the benefits. More recently widely disseminated revelations of Western powers' violence, hoarding, and propaganda show that the myth has recently run out of steam as the EU stagnates and the US fragments. Dangers abound in transition.
- Paul Feyerabend in *Against Method*[25] shows how Galileo defended his observations of our heliocentric solar system by constructing a complex, connected set of practices rather than a narrow scientific method. Galileo used rhetoric, propaganda, advertising, and epistemological tricks to defend his observations that threatened Church doctrine. The evolution from Galileo's observations to acceptance

of his findings moved through a complex ecology of channels using diverse methods.

- Jonathan Skinner has written that ecopoetics includes several diverse tendencies from pastoral to political to something like the environmental model I have framed:

For some readers, ecopoetics is the making and study of pastoral poetry, or poetry of wilderness and deep ecology. Or poetry that explores the human capacity for becoming animal, as well as humanity's ethically challenged relation to other animals. For others, it is poetry that confronts disasters and environmental injustices, including the difficulties and opportunities of urban environments. For yet others, ecopoetics is not a matter of theme, but of how certain poetic methods model ecological processes like complexity, non-linearity, feedback loops, and recycling. … Or how poetic experimentation complements scientific methods in extending a more reciprocal relation to alterity—ecopoetics as a 'poethics.' Or even how translation can diversify the 'monocrop' of a hegemonic language like English.[26]

These multiples between and within groups show evolution working at the social level to continually re-balance culture and aspirationally the human and non-human shares of the biosphere. Yet if any of Skinner's categories fails to reach beyond poetry into other disciplines and ways of thinking, they fail in their role as agents to slow climate change. But I think nearly all those categories, even the pastoral in some instances, the pastoral, engage other disciplines. Without extension into the other, ecopoetry can be no more than description with predictable effects on herds of cud chewers.

Skinner shows how "certain poetic methods model ecological processes like complexity, non-linearity, feedback loops, and recycling" to "all-of-making,"[26] a way of speaking about ecopoetics as all of poetry and echoing Elizabethan optimism about poetry's influence in Ben Jonson and Philip Sidney. How has ecopoetics scaled up to include multiple styles of writing and thinking? My reading of Skinner is that he wants to fashion the ecopoetic discourse through inclusion and connection, creating a taxonomy of ecopoetics rather than an evaluative hierarchy of good, better, best.

- Writing, in fact all human activity, involves ecology through inter-intra-action of humanity and the non-human. I have chosen to write about language writing because it often, although not always, dodges

> themes focused on surroundings, as in Hejinian's resource autobiography, but cannot avoid being engaged in fundamental ways with ecological connectivity. In general, poets and poems perform ecologically through a poetic ecosystem that includes editors, publishers, critics, readers, electronic media, and post offices. Poetry is a network.

While each poet represents a separate and perceptibly unique identity, each performs as part of a multiple, a composite, a componentized, and a temporal entity. Neither the group nor the individual alone is sufficient for existing, surviving, and thriving. Groups are adaptive and heterogeneous. A supportive ecosystem, a group process, and a responsive individual are all indispensable features of poetic life. The ecology of language-centered writing is barely noticed by its participants as visible in Hejinian's resource autobiography above that still twists itself into knots to avoid discussing poetry's environmental processes. Some acknowledge that facts of ecology scale to poetry as in Bob Grenier's "Northern New Hampshire."[27]

In changing the way English departments operated, the canon they taught, how writers focused, and how poetry was read and observed, language writers (as in Galileo's example) used format changes, nonstandard vocabulary, homophonic translation, permission to experiment with received language, multiple formats of publication, archives of recordings, training a new generation of English professors at several universities, and other tactics. The evolution of poetry was achieved through allying with other disciplines like philosophy, linguistics, and science that asserted the materiality of language and exposing the technical, performative, and communications mechanisms by which poetry projected to the reader. As George Hartley says in his seminal essay on language writing referring to Silliman's concepts outlined in *The New Sentence*,

> The gaps between sentences (the locus of tension or 'torque'), the visible seams, here take the place of the line break and draw our attention to the materiality of the words as words, not simply as transparent signifiers.[28]

This distinction was crucial to the inclusion of language writing in the academy. Hartley speaks about the channels as "visible seams" between sentences in the way environmentalism speaks about borderlands.

When a group is defined by exclusion, as was the case with language writing and most other poetry groups, outsiders don't see the myth-making group as representative of themselves. Initially, only a few poets identified as language-oriented writers, since it started as a regional practice in San Francisco. Connections built the group to include multiple writers in SF, DC, NY, and Toronto. *Roof* magazine published issues focused on each city's poets.[29] As language-centered writing developed, some critics, in an attempt to homogenize poetry, began to erase geographic boundaries by speaking about nearly all modern poetry as language-oriented even including writers with quite different politics like Ashbery and Berssenbrugge.

These examples from geopolitics, science, and two from poetry—ecopoetics and language writing—show how difficult and complicated it is to make connections that scale upward in a hierarchy from specific events like poem to a more inclusive general case like poetry in order to influence vested interests. How much work is required to change generalized biases! Difficult but possible. In some cases, the older ideas, such as religious bias against principles of evolution or poetry bias in favor of spirituality or immediacy, never disappear but are merely circumscribed by the newer framework. It's easy to see how residual concepts retain efficacy for individuals because every species defends its organism as well as its psychological and group isomorphs.

To figure out how ecological connections between biology, politics, and poetry can be engaged to scale upward to the general case of slowing climate change, understanding how different connections work helps prepare the reader for the extra energy required to move upward to an environmental view and ecological model. Mapping in the physical and conceptual senses makes the channels easier to navigate, because readers need to be convinced that this way of looking at things opens new avenues for their own work, their own success, happiness, and security.

The Green New Deal, for example, architected by Rhiana Gunn-Wright, Demond Drummer, and Zack Exley with support from economist Mariana Mazzucato and a variety of progressive public figures, shows how connecting jobs, job training, and clean energy production can enhance the possibilities of a society that is both fairer and greener. Claims about the Green New Deal as greenwashing capitalism and as a socialist monster are equally possible depending on how programs are implemented and divergent implementations are inevitable. But continuing production with new supply chains and processes must happen to reduce emissions.

In an example related to climate change, discussion of how to phase out fossil fuels in energy and politics is often avoided, but even Saudi Arabia is actively pursuing alternatives to fossil fuels.[30] Saudi scenario builders pay attention to how probabilistic connections work although with longer and more difficult travel time.

Poetry doesn't change simply by writing a poem; that's one step. Writing many poems and increasing awareness about the change through a variety of channels—print, online, readings, discussions, criticism—do far more to change poetry than any one poem. Not to reduce the value of a wonderful poem, but more complex efforts beyond the individual poem actually change assumptions. The programs of English Romantic poetry including poems and criticism, New Criticism poetry and poetics, NY School writing, publications, and educational institutions, US language writing, and ecopoetics are good examples of multidirectional influence.

Example 7: Connectors of Organism + Society = Self

Within any organism really large numbers of different pathways—vascular, biochemical, and electrical—connect organs, cells, and chemicals. In spite of all these components and connections that scale to the partite construction of the biosphere, I feel like I am a thing, even more specifically a person, even more specifically James Sherry, even more specifically here James Sherry the writer. What I am pointing to is the evolutionary advantage of thinking as a unit built on a questionable perception of wholeness and separateness,

a pile of shreds
on the heap of a hill.

Just how valid is my feeling of identifiable wholeness? Since I have already discussed the feedback from Graziano's attention schema as validating my sense of existence, I need to verify that an organism is a thing separate from other things. My genome scales to organism to ecosystem to biosphere, how can I be separate?

The way the biosphere is organized, there are a greater number of strong, multidirectional links within an organism than between organisms. The human brain has about 86 billion synapses (connectors) and a whale has three times that amount. While the individual extends into the air around them, into social and habitat space, links are less dense and less complex than those within a body whether you're a person or a tree.

Organisms may be identified by their density of linkages perceived by both outline and shading. This view works with both mainstream theories of how organisms are organized, the layer cake theory and the mechanism theory, as well as the more nuanced Local Maxima account of William Wimsatt.[31] In general, scientists view the organism as composed of several levels such as organ systems, organs, tissues, cells, organelles, molecules, atoms, and sub-atomic particles.

Poetries on the other hand have their own organizing principle around different interactions of language, social interaction, and memory. The organismic binary predicts more connections within a poem and within a group of poets than between poems and groups. In my experience, the language writers' group, the NY School group, and Iowa Writers' Workshop are organized by certain characteristics of style, by personal affiliation, and by the behavior of peers and predecessors. Individual writers within each group overlap somewhat and have individual characteristics of style. The institutions of poetry from universities to non-for-profit organizations to poetry clubs at the local level build group affiliations that have some hold on most of the writers who ever participated in them.

Poems have strong internal relationships by theme, logic, sound echoing, spacing, and other linkages within poems. The same linkages can be said to exist between poems, but connection that may be said to rhyme between poems is mostly arbitrary. Serial poems and some books of poetry can pay attention to a few linkages between poems, but rarely have affiliations as strong as within a single poem. The poetry world through poems and institutions mirrors the organismic world, a kind of scaling.

I discussed this hierarchic flow in the chapter "Scalable Syntax: Poetry Model of the Biosphere" where the poetry model scales through phonemes, poems, books, authors, editors, publishers, groups of writers, distributors, post offices, including paper mills and software engineers. I'd like to essentialize poems alone on the page like so many poets and critics before me, but environmentally aware poetry identifies how poems travel from concept to history to connect in the ecology of poetry.

The way people distinguish network connections differs from the statistical findings. This perceptual level dominates daily interactions across the network. People need to solve problems of race relations even though race cannot be well differentiated as a meaningful category in scientific terms even though some folks need to feel they're better than someone else. Leaders know society needs to solve problems of allocating resources between economic classes but they disagree how to distribute them:

- equally,
- according to need,
- according to ability to optimize them
- in a competitive marketplace or
- using several of these methods.

These multiple yet partial examples of connections at the conceptual, biotic, social, and ecosystem levels give the reader some idea of a tentative modeling of an environmental network of connection. Science and poetry tend to see the world through models because brains operate using those simplifications or schema which are kinds of models. Self and world are made of similar stuff. Thinking about self as an organism and social forces takes some of the conflict out of relationships. I continue to search for single answers to match my own misperceived singleness.

Socially, leadership often merges unnecessarily with control. Me Too and other women's groups have pointed out that they often lose control of their own boundaries through unequal power relations. Black Lives Matter proposes that Black women choose a perspective or work to integrate solutions for both race and gender. Regarding leadership, Cathy Cohen says:

> many of the young leaders in the Black Lives Matter movement recognize that the male charismatic leader, or the singular charismatic leader, is not the form of leadership that they adhere to or they going to put forth.[32]

In *Our Word is Our Weapon*, Subcommandante Marcos suggests a style of leadership that does not include taking power over others.[33]

Considering ideals, can we start each conversation with the presumption of equality and self-determination and then investigate? Crenshaw: without looking at intersections, people who have intersecting marginalized identities are left making (too) many compromises.[34] Police habitually assume that they have control over the boundaries of citizens in the name of the law, generalizing to all boundaries when in fact only certain conditions and laws allow police to cross a citizen's boundaries against their will.

When working alone and collaborating, the skin of a person is one of the primary boundaries. The environmental model shows the incredible complexity of the skin is synecdochical to the even more complex set of borders of the self. Myself and my friends wish it were simpler; I conflict with those who insist it be simpler.

The individual ranges beyond the skin socially through their family, gender, and community affiliations. On a larger scale, activities such as professions and hobbies extend identity. Then, nationality and other geographical connections are included in the thoughts we have about others that link us freely scaling to them as individuals and groups. Often, however, I think something about a person, but when I meet them, they seem very different than what I imagined. Our surroundings confuse us too. They look different when we're in them than they seem from far away. The smooth bed of tree tops becomes a frightening forest when we enter it. Many connections change with distance. The data is present for individuals, groups, and habitats, but its meaning is always relational, that is, dependent on how we are connected to each other and our surroundings. Donna Haraway:

> *When Species Meet* works by making connections, by trying to respond where curiosity and sometimes unexpected caring lead." And later "…mattering is always inside connections that demand and enable response, not bare calculation or ranking.[35]

Therefore, where I stop and you start is always conditional, situational, dependent on how we are connected, and often contentious regarding what and how we perceive. I may feel you're standing too close to me while you want at most that distance between us to confirm by feedback that you have my attention. The more anxious I am for my safety, the more distance between us I'd like and the more control I seek. Precarity varies with my distance from the cliff that dizzies me. In social situations like public transit or a party, I lose control of my borders. I become part of the habitat.

Models of environmental poetics (including ecopoetics and Black Mountain poetics), social ideas of poetry (language poetry and poets focused on group identity), and individual execution (Romantic, Confessional, NY School, assumptive individualism of specific identities) of poetic works all have different perspectives on how they connect within themselves and to others. They operate in writing, reading, publishing, and distributing ideas about their works. Their actions in the poetry world also create feedback loops that reinforce and undermine perspectives and actions. In the end, these poetry communities as an example of biologies and societies, and the composing selves inter-intra-act through the following matrix:

Perspectives			
	Person	Group	Habitat
Person	Person view of self	Group view of person	Habitat view of person
Group	Person view of group	Group view of self	Habitat view of group
Habitat	Person view of habitat	Group view of habitat	Habitat view of self
Actions			
	Person	Group	Habitat
Person	Person acts on self	Group acts on person	Habitat acts on person
Group	Person acts on group	Group acts on self	Habitat acts on group
Habitat	Person acts on habitat	Group acts on habitat	Habitat acts on self
Feedback			
	Person	Group	Habitat
Person	Person feedback to self	Group feedback to person	Habitat feedback to person
Group	Person feedback to group	Group feedback to self	Habitat feedback to group
Habitat	Person feedback to habitat	Group feedback to habitat	Habitat feedback to self

A perspective according to George Herbert Mead is "the world in its relationship to the individual and the individual in his relationship to the world."[36] Initial perceptions often produce different results than detailed investigation. Such perspectival analysis is often overlooked with perceptions assumed by individuals to scale readily to the results of thorough analysis.

Depending on your point of view, some, any, or all of these cells might contain physical analogs to the universe you inhabit. Not everyone can agree to the possibility that our collective surroundings feedback and scale down to a person. Not everyone agrees that a group can act on itself and evolve. For some of us, many of these cells are language constructions that have no analog in the world as they conceive of it, just like the possible amino acid combinations that only exist in this matrix. Yet for some, every cell is inhabited by real-world perspectives, actions, and feedback. In addition, each of these three matrices nests within the other when you take that point of view. Comparing perspectives and perceptions can be a useful environmental (top down and bottom up) and poetic (rhetorical) method.

The Condition & Hierarchy of Identification

This chapter presents the outlines of an inclusive hierarchy that resembles taxonomy as a more direct and accurate way to view the environmental model than the evaluative hierarchies of humanism and the organismic binary. I relate some of my own experiences changing my role by simply changing my schedule. I position identity to act like a fluid, as multiple iterations of self reform. I show examples like where poetry's social networks develop similarly to Darwin's finches' beaks to demonstrate scalability across disciplines. These ecologically described examples also reveal how people view themselves differently through refractive environmentalism than reflective humanism.

* * *

To identify the levels of the writing process that interact between poetry and ecology, to reconcile competing hierarchies—evaluative (organismic) and inclusive (environmental)—I will commit to a preference for solving problems of climate change with inclusive hierarchies based on the q-analysis developed by Ron Atkin in *Multidimensional Man*[1] and will show several examples of such inclusive hierarchies.

In this essay, however, I started by evaluating the roles that comprise and compromise, shade and shape my identities, appreciating the writer

J. Sherry, *Selfie*, https://doi.org/10.1007/978-981-19-4870-1_19

role more than other roles. In actually researching, writing, editing, and publicizing this essay, I included myself in the hierarchy of publication as well as the themes of the essay in order to play a convincing role. I frame this essay to present different styles poetry on the same level and then theories of poetry on the same level in an inclusive hierarchy of multiple dimensions.

In 2018, I volunteered to help get out the vote in a Congressional district, NY19, where I own a quarter-acre estate. Sometimes I am convinced by my role and feel good about it. Sometimes the role is a temporary means to an end and conflicts with my cherished writerly identity. I'm trying to identify as a writer most of the days I edit this manuscript, but frequently I see myself as the editor and then the publicist. In volunteering to make phone calls for the candidate, I take time away from the writer and become an agent of candidate Antonio Delgado.

Sometime a single role changes value when conditions change. I move from writing early in the day to making calls afternoons in NY19. Sometimes I raise poetry to high status. Sometimes I see it as a burden. Sometimes I don't think about poetry. My evaluation of the status of poetry in the world, for my group and for myself varies with my condition and the conditions around me. When I judge these roles against how I value myself as a whole, the evaluative hierarchy dominates.

My situation, however, involves more than how I evaluate myself and others. Several identity components reform as roles change. Which of my identity components come into play when I call prospective voters compared to when I play the poet compared to when I stand in line to get pay for groceries during the Covid-19 plague? What energies are stimulated; what emotions are engaged; what postures are taken? The engineer, too, operates differently in the design team, when reporting progress to management and upon arriving home to weed her garden. My patient, suffering citizen shuffles forward in line at the Department of Motor Vehicles. My earnest poet speaks at Columbia University differently than he speaks at the Poetry Project. The voice pitch and the vocabulary change through the themes addressed at different sites no matter how I try to hold my own. These changes in discourse are not so much evaluated as included as part of the situation/habitat where I am speaking.

Oh, no, you say, I am always myself. What I have to say to that comment doesn't play well in the polite company of these abstractions. Environmentally, there is no clear and distinct self and no single group identity, but rather a series of linked perceptions of self. Even the biosphere is a multiple based on perspectives that include the scientist's view, the corporate view, the householder's view, and the hunter's view. But the gentle reader knows that already. If you've read this far, you are prepared perhaps even eager to understand the multitudes you contain and recognize your merging into society even though Whitman was a racist. Even when I say that I'm not stable or satisfied, but reserve a core of self that sustains me, I might agree that is true at one scale, one level of graininess and extent, but not at others. While structures of the biosphere scale, our response to them changes. Consider what happens when you are fired from a job just after you signed a mortgage, were presented with a lawsuit from your spouse, and told you have a serious illness. Certainly, the core your identified self might crumble or crack with this kind of news and you'd need time to regroup.

Identity operates more like a fluid, layered, multi-dimensional surface (relevant metaphors abound) than a label of self like a museum display. You operate with many labels and senses of self. Self responds differently at different scales and groups differently. It scales, although, at times, it doesn't. While most people are more comfortable when they can speak about their identity as fixed, I would not think such stasis desirable in a constantly changing situation? Even the monk who tries his best to be one finds his roles as a monk many and varied, differing when praying than when serving the poor than gardening at the monastery than when remembering his evil twin. Does such an entity persist over time? Is he the same when confronted by three large, homeless men excited by the thought of his donation to their next meal? Is he the same when meeting Sheldon Adelson?

Although many institutions are built around a fixed identity, finding such a person is rare. How many movies depict a young, male, violent, drug-soaked gang member? Does such a static person persist over time even if there are people passing through violent stage at any one moment? Do some components change as the individual ages, gets clean, finds a job, meets his aunt on the street, and takes on responsibilities in family and community? Perhaps fixity is violence in a changing world.

Conditional roles are also complex since some are donned intentionally while others are imposed from the outside like when labels like gender, age, ability, class, and race are applied pejoratively. I imagine what it would be like if people only saw me in one layer, say religious heritage, and refused to acknowledge any of my other identity components. My identity would be not even of my own making. I might be thinking about lunch, my manuscript, and friends when an anti-Semite decides that that person that he sees in front of him is Jewish and that's all he sees. My life would be a great deal less entertaining for me. I'm a poet, publisher, contributor of a variety of social activities, voter, father, husband, friend, neighbor, shopper, retired computer jockey, old man with a short future.

Humanism on the third hand asks others to recognize primarily one's career role as doctor, poet, or electrician and encourages privacy about other components. When I mentioned my role in technology to a literature professor at a lunch in Bordeaux, he pretended not to hear me. Some people like to be treated that way—public components and private components. And of course, I don't want you to know everything about me—yuck. The matrix of views of nature predicts differences that are neither wrong nor right, but rather inclusive of comparative levels of value.

Condition extends beyond how people treat each other. The fireman behaves differently in a three-alarm house fire than when called out for a burning garbage can even if he wears the same outfit. Ladies who lunch (of which I am sometimes one) wear different outfits for shopping and cocktails. How do their points of view change? Condition also forces people into unpredictable roles such as when a gas explosion hits the massage parlor in Rodrigo Toscano's *Explosion Rocks Springfield*,[2] when the thief finds an old woman pinned under the rubble after an earthquake, when the poet attempts to continue reading to their audience while fire engines alarm the neighborhood, or the Dragon Dance is being performed in the alley, when the slave woman is released by UN peacekeepers into the hands of a born-again priest.

If you ask them, they recollect
their stalwart positions,
but what also occupies their minds
then in changing surroundings?

A person's various identities interact in communities like species in ecosystems and phrases in a poem—similar and different, scaling and metaphoric. The relationships between them act together and condition behavior. At another level, each individual can be described as a community/ecosystem in themselves, comprised of many organisms and acclimatized by many cultural, language preconditions. Such scalar, granular, similar, proportional, metaphoric connection, and perceived simultaneity show how selves change, are changed, changing, and changeable.

The patriarchal notion that a man must be himself, true to himself and inflexible in pursuit of the truth, strokes the perpetual hard-on of male-dominated nation states. Patriarchy rules even more harshly at the community level than for nations as in Orson Wells' *Touch of Evil* and more harshly still for the individual as in Franz Kafka's *Metamorphosis*.

The ability to describe this myth of fixed identity without causing cognitive dissonance for the reader implies to me that our language already admits the breakdown of structure where descriptions fit a thing but don't have a physical analog, what Wittgenstein calls spirit. Biological naturalism is built into language because language is built into the mind by biological processes that grow organisms able to reproduce. None of this should come as any surprise because we want to, as noted many times, feel unique and separate. The difficulty is how to acknowledge this scaling similarity in the way people talk about humanity, act toward each other and their surroundings. How can that be accomplished? How can culture reflect identities more accurately? How can the poetry invoking epiphany and momentous occasion also engage such conditionality?

What are the mechanisms for promoting a more complex view of environment? Shelley's "Mutability" represents one example of thinking about conditionality. Although he recognizes scaling: "It is the same!—For, be it joy or sorrow," he wants the "The path of its departure still [to be] free." I would say he's wrong to suggest "Man's yesterday may ne'er be like his morrow; / Nought may endure but Mutability."[3] Rather, many aspects of today persist tomorrow, similarly and differently. His insistence on finding a constant in mutability reiterates the Romantic error of self-identification that to slow climate change, we must correct.

Once an identity component is defined (man) and then another one (poet) and then a third (citizen), I need a taxonomy to variably perform myself in a society that constructs and uses these identities to accomplish living together at the scales of individual, social, and habitat. I also need to show identities that society can use identity to be less destructive of the surroundings without losing the improvements of civilization—longevity, health, freedom from privation, communication, mobility, etc.

Focusing on one role as my primary identity like a character in *Canterbury Tales*, I used to go to my job and prioritize the worker, subordinating the other layers of my identity. The poet, the critic, the publisher, the news junky, the cook, the citizen all became less important and little else occupied my mind while on the job. I was sullen about it and reassured myself that it was not *me*, but I spent nine or ten hours a day in this posture, doing this work and speaking with the appropriate vocabulary as a worker on the job. Leaving the office, I passed a group of children whose arguing triggered my parental identity. I had a teenage son at the time. "Those kids…" I muttered. I came home, subordinated the worker, and cranked up the poet and the cook and the lover of my wife Deborah Thomas who was thinking of herself as some other person altogether.

Peter and Rosemary Grant document the potential speed of evolution in *40 Years of Evolution: Darwin's Finches on Daphne Major Island* where the average beak length of the entire population of finches changes substantially year to year depending on rainfall which in turn changes available food sources that are accessible to one beak length more than another.[4] If climate changes quickly in one direction, birds and people, too, have the potential to change very quickly in one direction. Ecosystems are neither stable and ideal nor independent of human interaction. They don't change slowly or smoothly as this section heading attests. The sudden changes of ecosystems are well documented by Gould:

> Punctuation all the way up and down? The generalization and broader utility of punctuated equilibrium (in more than a metaphorical sense) at other elves of evolution, and for other disciplines in and outside the natural sciences.[5]

The interactions of individual, group, and surroundings together give us the best chance of seeing the world as it is. So, you can say, "for me in this situation" and "for us in this situation" and you can say, "for you in this situation I" and "for you in this situation we." Such formulations make more sense than saying "I can't speak at all about a person who I am not" because in many ways you are they. We know the other person is sad, even if we don't always know why. We know where a person is looking and can often decide why they are looking there. The other is neither opaque nor translucent, although we can make them opaque by ignoring everything in our surroundings except for the organismic binary. When thinking about this issue always keep in mind theory of mind as outlined above.

When their daily personal experience is dominated by stings and blows associated with one layer of identity, people lose sight of their environmental connections and revolve around one condition or role like policeman or poet. It will become increasingly difficult to reassess humanity's relationship to our habitats as ecosystems become increasingly hostile. The energy put into our survival as workers and poets masks the environmental model because that is the way the organismic binary operates. Queer identity may not be the best example, since it intentionally changes with role and appears decisively performative per writers like Judith Butler. But because queerness breaks the binary into many alternatives (Facebook as of this writing identifies 50 gender roles), examining how identity works for the queer individual with and for others emphasizes the complexity of environmentalism and the need for a non-evaluative taxonomy in our social interactions.

We need those identity cells to be well defined in order to see how they interact and link. Fiona Templeton:

> the you of my voice is both you, reader
> and you, prisoner
> the you of body
> for you I continue this scratching hand even as dark comes
> my world is small but I add it to yours[6]

Connections between cells operate with a performative component as companions and observers of each process. In order to proceed, it will be helpful to show how the traffic travels between the nodes.

Most people will understand the traffic between nodes that have direct connections like the traffic between the mother and child or the police and citizen. Others will easily follow the network traffic of a text from sources to writer to editor to publisher to distributor to reader. But many nodes are not directly connected and traffic between them does not flow smoothly: the relationship between poetry and technology is not direct. In another example, unless you know a publisher personally, how do you get your poetry published. Even if you know the publisher, they may have many reasons not to publish your book. Traffic flows in complex, indirect, and unpredictable channels between the poet at their desk and the publisher at theirs.

Looked at from an environmental perspective, however, there are additional factors. The poet may be a publisher himself. The poet may know a publisher or someone who works at a publisher. The poet may know a publisher in his small group of gender-specific writers, but not know a publisher of a major magazine where the poem would find its most receptive audience. The number of alternatives makes it difficult to shoot one selfie that represents the poet's relationship to the publisher and it's easy to shoot one's selfie in (or should I say of) the foot.

Drawing the relationships, the traffic flow can be mapped through the hierarchy of nodes. We can talk about direct connections (1-dimensional) where traffic between the poet and publisher flows smoothly. Then, we can talk about several kinds of indirect connections where there is no 1-dimensional connection between the poet and the publisher.

We can only draw three dimensions on this page, but the levels of connectivity are manageable in three dimensions. One poet (P1) is a white poet and has no activities other than poetry writing in the poetry world. Another poet (P2) is a Black poet who organizes and curates a reading series. P3 is a designer, editor, and publisher of a line of artist books. They do not identify as a poet or particularly about race and gender, but rather as an artist. P4 a professional editor curates a reading series and identifies as a poet. P5 is a Black publisher of a major poetry magazine who edits and also writes poetry. P6 is a publisher at a major publishing house and does not identify as a poet although her publishing house prints eight poetry books a year when she used to be the editor. Obviously, this is not a complete list, but I have tried to cover as many different levels of interest as needed to fill out a clarifying set of relationships. Gender isn't specified but will also play an important role. The roles are designated as follows:

R1—white poet
R2—black poet
R3—curator of reading series
R4—organizer of reading series
R5—editor
R6—designer
R7—publisher of poetry
R8—publisher including poetry

The set P of poets includes some of the poets we've been discussing. If a poet participates in a role, they are designated by a 1. If they do not participate in that role, they are designated with a 0. The matrix of six poetry world people looks like this.

(+)	*R1*	*R2*	*R3*	*R4*	*R5*	*R6*	*R7*	*R8*
P1	1	0	0	0	0	0	0	0
P2	0	1	1	1	0	0	0	0
P3	0	0	0	0	1	1	1	0
P4	1	0	1	0	1	0	0	0
P5	0	1	0	0	1	0	1	0
P6	0	0	0	0	1	0	0	1

Poets P2 and P3 have diverse roles and the most connections in the poetry world, but they are not connected by any role to each other. As a result, it is difficult for them to communicate and the traffic between them is complex, i.e., not direct. For them to have a conversation about what they share, they should include poet P5 who is both a poet and an editor/publisher in their conversation. Their conversation is affected by the fact that P5 is affiliated with a Black poetry magazine. While he is supportive of the work of P2, he is less interested in supporting P3, not antagonistic, just less supportive. As a result, the conversation even with all three of them together is less fruitful than it might have been had all three been identified with a Black group aesthetic. Individual proclivities and personality matches and mismatches determine how much they can get accomplished toward their mutual and individual goals.

Poets often find it easier to get their work published through their group affiliations than by directly going to a publisher or editor outside the group and relying on aesthetics alone. The Black poet (P2) has a direct connection to the Black editor-poet (P5). The poets who are editors also find it easier to publish through their connections to other editors as in role P4. When there is no connection as between P1 and P6, the traffic between them is indirect and slower, having to travel further to get to the desired destination in the same way that P2 and P3 need P5 to support their communication.

Goals about publication vary between poets and the editors. An awareness of who is writing new and potentially exciting poetry is important for all contemporary poetry editors. Personal relations both domestic and professional may go further than this simple role model. The poet, P1, who just writes poetry has a more difficult time getting work published and relies on an indirect and complex aesthetic to carry the traffic of their work. It would assist this poet to understand the environmental model, not just their binary of self and other.

Conversations at gatherings revolve around other poets, poetry, publishing, and a chance to speak like poets like to speak. Often an individual has difficulty talking about someone other than themselves when a severe attack of organismic binary dominates them whether due to prior anxieties, being excluded from a group or some combination. Another reason that poets gather together may be to meet people in roles that are similar to or different than their own. Having hosted many of these gatherings over the past 40 years, I have noticed how poets form groups in an ad hoc way at gatherings and have a hard time walking away from them due to inertia. People reform similar groups that were productive and pleasant for them at prior gatherings. Only certain individuals seek out those that they don't know.

The narrowness of the sets of affiliations and groups at gatherings has always been surprising. Comfort levels of conversations are greater among similar individuals. There is at a party, for example, often a closed group of the significant others related to poets. They speak quite freely compared to the guarded, topical conversations of the groups of poets who form highly adhesive connections around common themes like who is publishing where.

The organismic binary molds the groups at gatherings of poets as much as other affiliations of specific types such as race or style of poetry. The surroundings of the party often disappear for these groups. At these gatherings, party goers must both protect their organism and the needs of communication to disappear the venue and speak freely. It's not surprising then and therefore that people dedicated to their work often ignore the effect of their work on their environment. It is, I think, valid to generalize from this example—or should I call it a metaphor of the dynamics of environmental degradation.

What is not surprising is I am modeling this picture in a rather antiseptic simplification that doesn't show all the levels of detail that we all know complicate any gathering such as who slept with whom ten years ago and who resents whose success. And multiple levels of identity figure into behavior. One identity layer—race, gender, or even poetic style—is insufficient to designate an individual, but it can generate a party group through linking at one common level of several individuals.

The closer we are to another person, the more specific, multiple, and nuanced their identity layers appear. Anyone would be insulted to be designated solely as an Asian by their lover. If we're farther away from a person, we tend to sketch their identity more graphically and more along the lines of the sets above. This fact alone justifies including specific information about people (some of it physical), politics (some of it categorical), and aesthetics (some of it judgmental) in writing about someone.

Even in process-oriented writing like Flarf, details about individuals emerge. Part of the point of process-oriented writing is to use it to show that personal details always emerge in writing poetry that method is connected to identity. The connections established in the book *Motes* by Craig Dworkin[7] by overt imitation of the strategies of Grenier and Mac Low's influence on Anne Tardos in her process-oriented work *Nine*[8] but not in all her writing show different types and ease of connection and influence.

Even the most abstract or primarily musical writing links to psychological and social information about individuals. Heterogeneous strategies like language writing link the writer to the reader with multiple connectors. A matrix of techniques could be drawn in the same way as the role designations above. Linguistic strategies like the use of common, elevated, and complex speech and writing point to individual appetites, aesthetics, and politics. The similar shapes and frames of how writers approach language in poetry appear as a kind of "natural" form across many ways of thinking and being. Changes to our conditions also follow these same non-linear configurations.

Impact of Text & Environment on Another Self

This chapter discusses the effects of climate change on writing, individual, industrial, and social behavior, where they are similar and where they are different. The chapter shows how silos of knowledge prevent social change. The chapter shows how good and useful behavior now might not have the same effects later. It suggests that seeking correspondences between individuals and across social groups will become increasingly important as we understand how a society that operates ecologically can best support individuals, groups, and the biosphere. In this frame, the chapter discusses altruism and self-interest to show how ranges of behavior need to be accepted over demands for specific ways of acting.

* * *

Species' senses distort. Surroundings diffract events. People are repressed in some useful and some unpleasant ways by capitalism stimulating desire and religions asserting reality is illusory and humanity is fundamentally corrupt. Academic disciplines and even their scalability lock both productivity and thought into channels that are weakly connected to other disciplines and scales of knowledge. These isolated network segments and consequent environmental conditions propagate a field of conflict even as disciplinary specialization has created whole new fields of knowledge and wonderful inventions.

The target is to slow climate change, but will leaders in politics and academia give up how they control others to secure their power? Probably

J. Sherry, *Selfie*, https://doi.org/10.1007/978-981-19-4870-1_20

not easily; oligarchy will continue to operate in all institutions. Leadership by small groups is universal among animals. Will those in sciences and humanities who investigate the most accurate view of reality they can uncover open their disciplinary boundaries? I speculate that the best hope may be inclusion of cross-disciplinary studies, because academics are unlikely to give up even their limited power. *Selfie* was rejected by University of Iowa press because the editors can't sell inter-disciplinary books. The developed world, our knowledge, and our actions are siloed.

These politicians and researchers in science and humanities have figured out how to manipulate their surroundings to human advantage, but in the process destroyed many supportive multi-disciplinary habitats. Their narrow success resulted in a wider environmental failure. Environmentalism has four things to say about this topic:

Funny and inoffensive,
Funny and offensive,
Unfunny and inoffensive,
Unfunny and offensive.

I can learn more about the perspectives of power from this matrix of characterizations than I can from blaming investment managers and Catholics. But this matrix ignores levels of offensiveness that oppress, microaggressions that undermine, and it's somewhat annoying when I don't present an either/or choice that would support the organismic binary. The complexities go all the way down! Truth often upsets by exposing hidden motivations, therefore, if we only write texts that offend no one, how, then, will writing reveal any truth that changes assumptions or corrects imbalances?

If every layer of a person's identity, every act, must be considered "good" to avoid criticism for missteps, open societies won't last long because only those concealing their faults will survive. While some vehemently deny any chinks in their literary armor, some blame others to conceal their complicity. But if we include with each binary their connectors and characteristics, then the good is no longer only "what is aimed at" but also what travels successfully from individuals and groups through links to their surroundings. By noticing and facilitating such circulation, change becomes a priority as much as an inevitability and accommodating it a social value.

Acknowledging the complexity of traffic between points of view slows our responses to the speed endemic to life, one of the slowest forces in the biosphere—realization. Along with slow food, we shape slower interaction to replace snap judgments. We review first, second, and third thoughts to evaluate and exercise them. Sometimes intuition applies especially when immediate action must be effective. But building aesthetic and political positions from environmental models instead of the target of the moment from unexamined assumptions helps me accept that another individual's method may be quite different from mine. Part of my group's method accepts and examines difference.

Differences in poetic style, as discussed above, result in exclusion of poets from reading lists. These differences express methods of protecting current opinion and access to available poetry spaces such as books, magazines, readings, discussions, prizes, academic chairs, and even the physical microphone. David Levi Strauss posted a good summary of the "poetry wars" between the magazine editors of *Poetry Flash* and language writers in the mid-1980s.[1] Some of my closest colleagues in the poetry community actively attack each other sometimes for personal reasons, sometimes for stylistic differences. Canons taught at University of Iowa Writing Center differ from those taught at Columbia University or University of Pennsylvania. Personal differences and stylistic differences create dispute and hardening of borders. Fifty years from now will Hejinian and Jorie Graham be taught as irreconcilable poetries? Will Ashbery be taught as part of the same tendency as O'Hara?

Darwin and Wallace realized similar conclusions but by quite different methods and with quite different outcomes as far as public acceptance. Wallace's ethics was based more on the hope that humanity could improve itself while Darwin was more interested in questions of chance and will in moral issues[2] (Juan Manuel Rodríguez Caso). In both cases, evolutionary theory was central to the fragmentation of moral leanings, first in Victorian England and then spreading throughout the world as anthropological studies framed multiple views of behavior depending increasingly on the location of each group. Cultures without electricity differ from each other although they share structures, correspondences, and isomorphisms like individual identity, kinship relations, and perspective on their surroundings. Claude Levi-Strauss:

> The first step is the definition of the constituent units of an institution; these are conceptually equivalent to the phonemes and morphemes of a language and, therefore, comparable cross-culturally. Once the various

> aspects of a culture have been reduced to their cultural elements, relationships of opposition and correlation and permutation and transformation among these elements can be defined. Homologies between institutions within the same society or among various societies can be explained, not in terms of a mechanical causality, but rather in dialectical terms. Correspondences or isomorphisms should be sought, not between empirical data pertaining to different institutions, but between systematized forms or models, which are abstracted on different levels and which can be compared either intra- or cross-culturally.[3]

Like Linnaeus, Claude Levi-Strauss emphasizes form (morphology) and "the primacy of relations over entities, and on the search for constant relationships among phenomena at the most abstract level."[4]

Diverse evolutionists present a range of ecosystem interaction traveling along multiple channels to theories, each supported by many facts. Both Russian ideas of scarcity and Western theorists' love of tropic lushness are situated within ecosystems of thought derived from where they gather their data. The differences in contemporary evolutionary views again reinforce the notion that environmental positions are multiple with common elements that religions and ontology call essence. I suggest that essence displays characteristics of an isomorph of the relations between self, group, place, and their connectors more than a characteristic of their entities. Just as climate does not progress in only one direction but vibrates within ranges, evolution on a different scale tends to operate within ranges and only goes on one direction for a long time when climate and internal conditions change in that direction. Research continues to change views of evolution through the play of time in a plastic biosphere.

Multiple principles and behaviors operate through a diversity of texts, individuals, groups, and habitats and also through channels of multi-level selection aka group and kin selection. Multi-level selection demonstrates how altruism arises when individual organisms compete for survival. The argument between kin and group selection aside, social behaviors like altruism can be attributed to group/kin selection. Jos Kramer and Joël Meunier:

> Processes such as conformance transmission and especially the formation and enforcement of norms can create extreme phenotypic differences among groups and uniformity within groups—the raw material upon which multi-level selection acts.[5]

Within groups, selfish behavior defeats altruistic behavior because altruistic behavior requires that members of the group spend time, energy, and risk on behalf of others distracting from their ability to produce and build advantage. Between groups, however, altruistic groups tend to dominate over selfish groups. In a third model, Wilson and Omar Tonsi Eldakar: altruism and selfishness are linked.[6] So rather than a single perspective again we see a complex structure of behavior where it becomes increasingly apparent that working together (across groups for engineering solutions and across disciplines for ways to reimage the biosphere) limits global warming, improving chances for species survival.

Poetic efforts to assert altruistic behavior along with individual achievement and other group tendencies will encourage those behaviors among leaders and help avoid the total collapse of civil society. Poetry about self dominates many readers' ideas of the art and remains popular. A study of the comparative popularity of poetic foci would clarify how much this statement rather than quality of technique or accuracy of ideas attracts readers' attention.

Consider, too, the many avant garde, experimental, innovative, and investigative writing practices and forms of the past hundred years. Many are successful: Imagism, Dada, Futurism, Black Mountain, Language Writing, and many more began with group affiliations and then separated into individual poets through value assigned by the commercial and academic acceptance criteria of capitalist individualism. How hard the corporate poetry institutions have tried to separate Eliot from Pound despite their collaboration on *The Wasteland*, to distinguish Mayakovsky from Anna Akhmatova—two voices speaking for the Russian people—and to alienate named groups of poets like Surrealists, Objectivists, NY School, and Language Writing from named individual poets.

Different approaches to evolution appear within poems from Alexie Sherman's ironic Native American "Evolution"[7] compared to James Clerk Maxwell's ascerbic, anti-Darwinian "Molecular Evolution."[8] The diversity of organisms, points of view, and options for engaging with words connected through syntax permits more latitude than any creed of dogmatic phrasing, repetitive form, and predictable ideas accepted by all. Yet all are able to be compared. But rather than comparing these texts to evaluate their worth, the environmental model as I'm framing it includes them, compares them, and encourages diversity with categorization to understand how they add to the possibilities of actions and the body of knowledge about the evolution of and in writing. (Actually, I found few poems directly addressing evolution per se.)

If poets and readers accept ranges of practice in ecology and poetry within a loose confederation of values, they allow for diverse poems, shapes, and styles of despair and delight. This political condition arises from a dynamic set of intellectual disciplines, social roles, and care for our surroundings. Their reading of ecopoetics may be as variable and as fruitful as their relationship to the complexities of gender identity.

Threats from climate change tend to drive people inward. Communities, ideologies, and nations harden their borders against each other. At the same time, social media networks reconnect many by establishing self-reinforcing subnets of "friends" and followers. People who use the internet reorient allegiances to balance their needs with those of machines in many ways cutting out surroundings. The publishers of *Selfie* wanted me to change its title so it would be more easily found by search engines. I changed it from *Selfie: Poetry & Ecology* to *Selfie: Poetry, Social Change & Ecological Connection* to make it less catchy but to capture the word "connection" for machine reading.

At the same time, these lubricated corporate information and supply chains permit environmentalists to spread awareness of our surroundings to all nodes of their network, to change the habits of people who are listening.

Fast Rivers & Cash Flow
Preference rises as voice
against the massive
arms of dogma.
Workers stream into conflict
over difference. For centuries
they blush to ask for food
from leaders who exclude
them for profit
The ruling classes claim
that the world does not exist
and we are nothing
while they extract.
Balance and extremes prevail.

The laws against discrimination passed in the US in my youth and the political independence of many colonized peoples have not really changed oligarchic behavior. Political dependence has been replaced by financial dependence as nation states share control with corporate boards and all depend on calculation by digital computers. Large swathes of humanity have been reconnected, making cyberspace its habitat and playing into the hands of a new group of oligarchs from Elon Musk to Jack Ma to Jeff Bezos.

Good Luck
Minority groups seek to change
how majority groups
pressured by power
act toward them in the network.
Each hopes for better balance
by hardening borders.

Protecting the organism and hence the self, its identities, connections, and assumptions seem primary concerns for almost most individuals. All seek some combination of freedom, support, and protection for themselves, family, society (corporate, state), and habitat. Even animals would prefer freedom of action with freedom from privation. But such privilege requires an underclass of support. With an overflowing human population, this desire poisons the planet and is endemic to life's scalable dialectic. Progressives in my youth prioritized truth at all costs. Now progressive youth avoid harm at all costs to protect precarious individuals and groups. Culture changes how we view and treat our surroundings.

No single binary of gender, race opposition, class war, and aspersions of ability circumscribes entire persons dependent on multiples over time and location, connectivity, situation, and consequent mutability. Then, we have a basis to say, not everything I do and believe operates in the binary sense but rather interoperates through components that fall in a range of workable solutions to a transport problem ("trains, planes and automobiles"), an ontological problem ("are you or have you ever been?"), or a poetic problem ("poetry been very very good to me" (a group of language writers riffing on Saturday Night Live)).

Can humans all dispense with the organismic binary and allow a range of less harmful security measures to guide our action? Probably not, but rebalancing the overweighted position of individualism might help increase the attractions of shared experience of poetry. As much as embattled poets would like it, not every action and value must be defended as if it were their body. Accepting rangy behavior is critical to environmental success for our species and, if looked at fearlessly, vital to social and poetic dynamics. Accepting personal, political, and environmental positions as contestable builds general consent and furthers our poetic, political, and environmental survival.

At another scale, every layer of my or my groups' identities acts with similar rangy properties. Collectively group connections and identifiable actions create a huge range of viable behavior. Language writers, for example, work as academics, business executives, techies, sales people, public relations writers, novelists, psychologists, social workers, and marginal free-lancers. Moral binaries don't support this complex set of conditional and partial indications but rather reinforce more binary behavior. While binaries persist because of the omnipresent and indispensable organism, engineers, poets, and activists need to look at ranges of behavior in more detail when working on environmental culture and industry because it seems quite easy to fall into destructive habits associated with the necessary exploitation of our surroundings required for survival. Mutual aid produces some non-binary combinations of apparatus and action that align with other non-binary inclinations in the emerging environmental culture. The focus on individual to group relations seems of vital ethical concern.

Individuals suffer when their interests and opinions differ from their identity groups. My affiliation with ecology has alienated many political dualists' affections. The conflict between Chelsea Manning and the US security state resulted in Manning's incarceration and continued oppression. At the end of his life, Malcolm X began to see race in global terms that conflicted with the Nation of Islam's group bias. Instead of allowing him to continue to speak out as a globalist, the Nation of Islam assassinated him. Erin Brockovich worked for many years for justice against Pacific Gas and Electric, a company that declared bankruptcy in 2019 as a result of failing to maintain its own infrastructure. Short-term, self-interested behavior caused the collapse of PG&E's infrastructure in the California wildfires of 2018 and 2019.

Poets and groups constantly vying to protect their exteriors and posteriors often neglect their primary purpose. Slowing global warming may in a similar way profoundly stress standards of civilization in developed countries. Addressing climate problems early and broadly will be significantly less expensive and cause less harm, but will cause financial disruption. Poets imitating those defended binaries may appeal to many readers but can be read as to threats to species survival.

Notes

Introduction

1. "Poetry and Abstract Thought." Tr. Charles Guenther. *The Kenyon Review,* Spring, Vol. 16, No. 2, Spring, 1954. 218.
2. *ABC of Reading*. Faber, 1934. 36.
3. "Edgar Allan Poe: His Life and Works." Trans. H. Curwen. *The Works of Edgar Allan Poe*. John Camden Hotten, 1873. www.eapoe.org/papers/misc1851/1873000m.htm. 5 February 2022.
4. "In Memory of W.B. Yeats." *Collected Poems*, ed. Edward Mendelson. Vintage, 1991. 247.
5. "Preface." *Lyrical Ballads*. Bartleby.com/39. 28 December 2021.
6. "Man Carrying Thing." *The Collected Poems of Wallace Stevens*. Knopf, 1954. 350.
7. "Poetry Is Life Distilled" quoted from Brooks and the title of her biography by Christine M. Hill. Enslow Publishing, 2008. 128p.
8. "This Was a Poet—It Is That." *The Poems of Emily Dickinson*. #446. Ed. R.W. Franklin. Belnap, 1999. 206.
9. "The Rejection of Closure." *The Language of Inquiry*. University of California Press, 2000. 40–58.
10. *The Combustion Cycle*. Roof Books, 2021. 512 & Epigraph.
11. *Wounds of Passion: A Writing Life*. Holt, 1999. 288p.
12. "U.S. Global Change Research Program." *The Climate Report*. Melville House, 2019. 13.

J. Sherry, *Selfie*, https://doi.org/10.1007/978-981-19-4870-1

13. Alice Hill & Leonardo Martinez-Diaz. "Adapt or Perish." *Foreign Affairs*, Vol. 99, No. 1. 115.
14. "Nature." poets.org/poem/nature-1. 5 February 2022.
15. University of Oklahoma Press, 1955. 246p.
16. *Canada's National Observer*. www.nationalobserver.com/2019/08/01/news/no-republican-strategist-didnt-invent-term-climate-change. 6 January 2022.
17. Among the partial, problematic, and misinformed definitions of language writing, George Hartley's rendition can be a starting point if readers have no background or knowledge of these poetry groups: www.writing.upenn.edu/~afilreis/88/hartley.html. 22 November 2020.
18. John Shoptaw. www.poetryfoundation.org/poetrymagazine/articles/70299/why-ecopoetry. 12 December 2021
19. University of Georgia Press, 2012. 435.
20. Nightboat Books, 2014. 190p.
21. www.ceciliavicuna.com/performances. 26 December 2021.
22. *This Magazine*, #7. This Press, 1976.
23. "Selfie." *OED*. 2021.
24. *PLoS One*. 2018; 13(10): www.ncbi.nlm.nih.gov/pmc/articles/PMC6201881/. 8 January 2022.
25. "The Effect of Climate on Human Behavior in History." *Transactions of the Kansas Academy of Science*, Vol. 46, April 10, 1943. 33–51.
26. www.science.org.au/learning/general-audience/science-climate-change/2-how-has-climate-changed. 12/29/2021.
27. Sabine L. Perch-Nielsen, Michèle B. Bättig, and Dieter Imboden. "Exploring the link between climate change and migration." *Climatic Change* 91(3–4). Permanent Link: http://hdl.handle.net/20.500.11850/1455. 7 February 2022.
28. *Report on the Impact of Climate Change on Migration*. Joe Biden's White House Report. October, 2021. https://reliefweb.int/report/world/report-impact-climate-change-migration-october-2021. 22 November 2021.
29. Plato. *The Republic*. Book VII. Trans. Allan Bloom. Basic Books, 1968.
30. *Time's Arrow, Time's Cycle*. Harvard University Press, 1988. 84.
31. "Poetry and Grammar." *Lectures in America*. Beacon Press, 1957. 223.

The Particles

1. Simone de Beauvoir, *The Second Sex*. Jonathan Cape, 1956. 13.
2. Julie Foertsch and Morton Ann Gernsbacher. "IN SEARCH OF GENDER NEUTRALITY: Is Singular *They* a Cognitively Efficient Substitute for Generic *He*?." Psychol Sci. 1997 March; 8(2): 106–111. https://doi.org/10.1111/j.1467-9280.1997.tb00691.x. 7 February 2022. www.ncbi.nlm.nih.gov/pmc/articles/PMC4293036/.
3. Modern Library, 1998. 689p.
4. Bloomsbury Press, 2011. 321p.
5. "On the Role of the Ventromedial Prefrontal Cortex in Self-Processing: The Valuation Hypothesis." *Frontiers in Human Neuroscience*, 2013; 7: 372. https://doi.org/10.3389/fnhum.2013.00372. 7 February 2022.
6. *Both Poems*. Roof Books, 2011. 11.
7. *The Origin of Species*. Op. Cit.
8. "The Imperative of Self-Reference in a Theoretical Framework for Biology." www.academia.edu/50664004/The_imperative_of_self_reference_in_a_theoretical_framework_for_biology September 11, 2021. 7 February 2022.
9. Premack & Woodruff. *The Behavioral and Brain Sciences*, Vol. 49, 1978. 515–526.
10. Perseus, 2002. 58.
11. "Self." *Echolocation*. Roof, 2018. 12.
12. "The Origins of Social Categorization." *Trends Cogn Sci*., Vol. 21, No. 7, July 2017. 556–568.
13. *Ojos de la arena. ejercicios de desclasificación*, tr. Erickinson Bezerra. ACCI Ediciones, 2020. 44.
14. "The Meaning(s) of Declassification in Antonio García Gutiérrez." www.academia.edu/50231535/The_meaning_s_of_the_declassification_in_Antonio_Garc%C3%ADa_Guti%C3%A9rrez?email_work_card=thumbnail. 12 October 2021.
15. *Flarf: An Anthology of Flarf*. Eds. Drew Gardner, Nada Gordon, et al. Edge Books, 2017. 288p.
16. "Evolution of Human Vocal Production." *Current Biology*, Vol. 18, No. 11. R457. www.cell.com/current-biology/pdf/S0960-9822(08)00371-0.pdf. 7 February 2022.
17. The Figures,1980. 98p.

18. "Jubilate Agno." *The Poetical Works of Christopher Smart, I: Jubilate Agno.* Ed. Karina Williamson. Clarendon, 1980. 87.
19. Ed. Tom Jones. Princeton University Press, 2016. Section III.
20. *Phylon.* (1945) VI, vol. 3. 249.
21. "Darwin's'Extreme' Imperfection?" *Evo Edu Outreach*, Vol. 2. 84–89. 2009.
22. *The Structure of Evolutionary Theory*. Belnap, 2002. 1433p.
23. *Imperfezione*. Raffaello Cortina Editore, 2019. 7.
24. I remain skeptical that Berra said all these things attributed to him, but I'm going with the flow.
25. "Poetic Citizenship and Negative Dialectics." *topsy-turvy*. University of Chicago Press, 2020.

Nouns & Things: Changing Their Climate

1. *Lectures in America*. Beacon Press, 1957. 231.
2. Karen Barad. "Posthumanist Performativity: Toward an Understanding of How Matter Comes to Matter." *Signs: Journal of Women in Culture and Society*, vol. 28, no. 3, Spring 2003. University of Chicago Press, 2003. 802.
3. public.oed.com/about/. 7 February 2022: 600,000 total words about half nouns. "Nouns Make Up the Largest Class of Words in Most Languages, Including English." www.merriam-webster.com/dictionary/noun. "The Second Edition of the 20-volume Oxford English Dictionary, published in 1989, contains full entries for 171,476 words in current use, and 47,156 obsolete words. To this may be added around 9,500 derivative words included as subentries. Over half of these words are nouns, about a quarter adjectives, and about a seventh verbs;..." www.lexico.com/en/explore/how-many-words-are-there-in-the-english-language. 7 February 2022.
4. Peter Godfrey-Smith. *Other Minds*. FSG, 2016. 74.
5. *Critique of Pure Reason*. Cambridge University Press, 1999. Bxxvi–xxvii.
6. Oxford University Press, 2013. 288p.
7. For details about this idea and further discussion of indeterminacy versus uncertainty, see *Meeting the Universe Halfway*. Duke University Press, 2007. Appendix C.

8. Random House, 2002. 259.
9. Weile Chen et al. "Root Morphology and Mycorrhizal Symbioses Together Shape Nutrient Foraging Strategies of Temperate Trees" (PNAS August 2, 2016 113 (31) 8741–8746; first published July 18, 2016. https://doi.org/10.1073/pnas.1601006113.
10. "Sonnet to Evening: Written under a tree in the woods of St. Amand, in Flanders". *Poems*. J. Bell, 1791. 175.
11. *The Biosphere*. Copernicus, 1998. 60.
12. *Op. Cit.* 151.
13. Ibid. 150.
14. *Conjure*. Wesleyan, 2020. 96.
15. *Op. Cit.* 33.
16. *Lectures in America*. Op. Cit. 231.
17. "Projective Verse." *Collected Prose*. Eds. Don Allen & Benjamin Friedlander. University of California Press, 1997. 247.
18. *Fast Speaking Woman and other Chants*. City Lights, 1975. 76p.
19. Moore. *Op. Cit.*
20. *Ecology without Nature*. Harvard, 2007. 264p.
21. *Op. Cit.* 35.
22. *Ibid.* 28.
23. *Leaves of Grass*. Library of America, 2018.
24. *Reassembling the Social*. Oxford University Press, 2007. 110.
25. *Ibid.* 4,5.
26. *Science in Action*. Harvard, 1987. 180.
27. Moore. *Op. Cit.* 27.
28. Barad. *Op. Cit.* 144.
29. "On actor-network theory: A few clarifications." *Soziale Welt, 47. Jahrg., H. 4*, 1996, 369. www.jstor.org/stable/40878163?origin=JSTOR-pdf. 16 September 2020.
30. David Harvey. *The Nature of Environment: The Dialectics of Social and Environmental Change*. Presented at Havens Center of the University of Wisconsin, 1991. 20. http://socialistregister.com/index.php/srv/article/view/5621/2519. 15 February 2022.

Verb Solutions Adapt

1. Godfrey-Smith. *Op. Cit.* 24.
2. Ibid. 81.

3. "Let's Talk Climate Change: A How-to Guide." www.nature.org/en-us/what-we-do/our-priorities/tackle-climate-change/climate-change-stories/can-we-talk-climate/. 12 December 2019.
4. For more detail see James Sherry. *Oops! Environmental Poetics*. "Part 1: Business Process Interruption and Risk Homeostasis: Analysis for Technology Solutions to Regulatory Impact." BlazeVox, 2013. 185–189.
5. Moore. *Op. Cit.* 24.
6. Barad. *Op. Cit.* 146.
7. "6 + 5 pieces from Webern (2015, 2017). *Scratches*. if p then q, 2019.
8. *Paradise Lost*. Book V. Penguin Classics, 2012.
9. *Collected Works*, ed. Jenny Penberthy. University of California Press, 2002. 194.
10. Stein. Op. Cit. 211.
11. *Ibid.* 223.
12. *Museum of Capitalism*. Talisman House, 2021. 149p.
13. Stein. *Op. Cit.* 230.
14. *Judgment Under Uncertainty: Heuristics and Biases*. Ed. Kahneman, et al. Cambridge University Press, 1982. 456p.
15. *Ibid.*
16. *Ibid.* 454.
17. *Darwin's Cathedral*. University of Chicago Press, 2002. 228.
18. *Predictably Irrational*. HarperCollins, 2009. 400p.
19. *Nudge*. Yale University Press, 2008. 384p.
20. Kahneman. *Op. Cit.*
21. *The Climate Report*. U.S. Global Change Research Program. Melville House, 2018. 165.
22. *Ibid.* 166.
23. "Second Avenue." *The Collected Poems of Frank O'Hara*, ed. Donald Allen. University of California Press, 1995.
24. "Chronic Meanings." *Virtual Reality*. Roof, 2019.
25. *The Climate Report*. Op. Cit. 170.
26. *Ibid.*

Borderlands: Am I I & They? Knowing Without Or

1. Gloria Anzaldúa. *La Frontera/Borderlands*. Aunt Lute Books, 1999. 288p.
2. Environmentalism speaks only of an increased probability that people with power over polluting processes recognize that places, things, and selves interpenetrate, since pressures on individual organisms and groups drive protective, differentiating postures where the outside, insofar as it exists, might be pretty nice but still largely unappreciated as continuing and extending the individual.
3. Smith. *Op. Cit.* 84,5 and 104,5.
4. "How Phantom Limbs Explain Consciousness." *The Atlantic*. February 11, 2016.
5. *Ibid.*
6. "Mending Wall." The Poetry of Robert Frost: The Collected Poems, Complete and Unabridged. Holt, Henry & Company, Inc., 1955. 33.
7. *The Secret History of the Dividing Line*. Telephone Books, 1978. 2.
8. Paraphrase from Bruno Latour, *Down to Earth*. Polity Press, 2018. 93.
9. Herman Melville. *Moby Dick*, "The Chapel."
10. *The Condition of Postmodernity*. Blackwell, 1989. 271.
11. *Ibid.* 27.
12. Henry Kissinger, *World Order*. Penguin, 2014. 25.
13. Jakob von Uexküll. *A Foray into the World of Animals and Humans*. University of Minnesota Press, 2010. 54–63.
14. I have modified some phrases from Latour. *Op. Cit.*
15. Zhao Si. "Green Love." *Urban Poetry from China*, eds. Huang Fan et al. Roof Books, 2021. 370.
16. Audre Lorde. "After Images." The Collected Poems of Audre Lorde. W. W. Norton, 1997.
17. Amiri Baraka. "Short Speech to My Friends." *African American Poetry*, ed. Kevin Young. The Library of America, 2020. 393.
18. Quoted in Richard Wright. "Racial Bias in Negro Poetry." *Poetry* 53:6. (March, 1944): 328.
19. Fred Moten and Stefano Harney. The Undercommons: Fugitive Planning & Black Study. Autonomedia/Minor Compositions, 2013.

20. Marsilius of Padua. Defensor Pacis. Internet Medieval Sourcebook. Trans. Oliver J. Thatcher, 1905. source-books.fordham.edu/sbook.asp. 13 February 2022.
21. Cure. "This. Here and Now. With You."
22. Myung Mi Kim. "[accumulation of land]" Penury. Omnidawn, 2009.
23. Cambridge University Press, 2011. 21, 22.
24. *Ecological Communication.* Tr. John Bednarz Jr. Polity Press, 1989. 44–51.
25. *Goodnight, Marie, May God Have Mercy on Your Soul.* Roof, 2017. 56.

Syntax or How I Become What I Seem

1. Steve McCaffery and bpNichol, *Rational Geomancy: The Kids of the Book-Machine. The Collected Research Reports of the Toronto Research Group 1973–1982.* Talon Books, 1992. 99.
2. *Gender Trouble.* Routledge, 1990. 15.
3. *Topics in Walpiri Grammar.* www.ai.mit.edu/projects/dm/theses/nash80.pdf. 15 October 2021.
4. "How Early Hormones Shape Gender Development." Curr Opin Behav Sci. 2016 February; 7: 53–60. https://doi.org/10.1016/j.cobeha.2015.11.011.
5. Lisa M. Thurston, William V. Holt and Alireza Fazeli. *Principles of Gender Specific Medicine.* Academic Press, 2017. 251–267.
6. This Press, 1974.
7. "just lookin just lookin" *My Hands to Mutant Solidarities.* Dry Imager Press, 2020.
8. Alexander Pope. "An Essay on Man". *The Poems of Alexander Pope,* ed. John Butt. Yale University Press, 1966. 507.
9. *Quanundrum.* Roof, 2021.
10. Rootsaction.org. 15 June 2018. http://act.rootsaction.org/o/6503/t/0/blastContent.jsp?email_blast_KEY%E2%80%89=%E2%80%89213,802. 8 February 2022. Also, Mohamad Bazzi. "The troubling overlap between Jared Kushner's business interests and US foreign policy." *The Guardian.* 8 July 2019.

11. John Laitner, et al. "Incorporating Behavioural, Social, and Organizational Phenomena in the Assessment of Climate Change Mitigation Options." Jochem et al. *Society, Behaviour, and Climate Change Mitigation*. Springer Netherlands, 2000. 1.
12. *Leaves of Grass*. 1891. Op. Cit.
13. www.awakin.org/read/view.php?tid=2252, 18 May 2019.
14. www.tibethouse.jp/about/buddhism/text/pdfs/Bodhisattvas_way_English.pdf. 3 March 2020.
15. *Origin*. Op. Cit. 213.
16. Scott F. Gilbert et al. "A Symbiotic View of Life: We Have Never Been Individuals." *The Quarterly Review of Biology*, vol. 87, no. 4 (December 2012), 325–341. www.jstor.org/stable/10.1086/668166. 24 July 2019.
17. "Of Experience." Penguin, 1993.
18. *Lyric Sexology*. Roof, 2016. 3.
19. Virgil. *The Aeneid*. Tr. Robert Fagels. Penguin, 2006. 6.
20. John Donne. "The Canonization." www.poetryfoundation.org/poems/44097/the-canonization. 29 March 2020.
21. Brenda Iijima. *Bionic Communality*. Roof, 2021. 56.
21. Paraphrase from *Consciousness and the Social Brain*. Oxford University Press, 2013. 25, 26.

Notes: Social Groups Scale: Identities & Connections

1. Vintage Books, 1993. 97.
2. Roof, 2021. 81.
3. Cheryl Crazy Bull and Justin Guillory. "Revolution in Higher Education." *Daedalus*, Vol. 147, No. 2, The MIT Press, Spring 2018. 97.
4. *Journal of Theoretical Biology*, Vol. 14, No. 3. 255–74. https://doi.org/10.1016/0022-5193(67)90079-3. PMID 11,541,392, 8 August 2022.
5. www.nih.gov/news-events/news-releases/nih-human-microbiome-project-defines-normal-bacterial-makeup-body, NIH News release. 13 June 2012.
6. *This View of Life: Completing the Darwinian Revolution*. Pantheon, 2019. 26.
7. *Ibid*. 9.

8. Rothenberg. *Op. Cit.* 320p.
9. Charles Darwin. *The Descent of Man.* Penguin, 2004. 157–158.
10. *Mutual Aid: A Factor of Evolution.* "Introduction." Retrieved on February 16th, 2009 from dwardmac.pitzer.edu, proofread source retrieved on August 5th, 2019, from revoltlib.com, 1902. 5.
11. Eds. Matthew Hofer & Michael Golston. New Mexico University Press, 2019. 440p.
12. New Directions, 2007. 160p.
13. Wilson. *Darwin's Cathedral.* Op. Cit. 28.
14. *Ibid.* 43.
15. *Becoming Undone: Darwinian Reflections on Life, Politics, and Art.* Duke University Press, 2011. 280p.
16. *Evolution and Culture.* Eds. Sahlins & Elman R. Service. University of Michigan, 1988. 133p.
17. "Evolutionary Epistemology." *The Philosophy of Karl R. Popper*, ed. P. A. Schilpp. Open Court,1974. 412–463.
18. Wilson. *Op. Cit.* 19.
19. "In the Corners of Our Lives," tr. Chang Fen-ling. *Urban Poetry from China*, eds. Huang Fan et al. Roof Books, 2021. 46.
20. *Proceedings of the National Academy of Sciences.* A study of 51 languages. www.pnas.org/content/117/5/2347. 1 April 2020.
21. Matthew J. Nelson et al. "Neurophysiological Dynamics of Phrase-Structure Building During Sentence Processing." www.pnas.org/content/pnas/114/18/E3669.full.pdf. 31 March 2020.
22. www.itproportal.com/features/how-to-predict-the-lifespan-of-a-company-in-one-simple-measure/. 14 February 2017.
23. Moretti. Verso, 2007, 81.
24. "Fight for Your Alienation: The Fantasy of Employability and the Ironic Struggle for Self-Exploitation." *Ephemera: Theory & Politics in Organizations*, Vol. 13, No. 4, 2013. 785–807.
25. Thomas Piketty. *Capital in the 21st Century.* Belknap Press, 2017. 816.
26. *The Stranger.* Vintage Books, 1958.
27. My summary of Pasolini's. *Saint Paul.* Verso, 2014.
28. Biden. www.npr.org/2019/06/19/734103488/democrats-blast-biden-for-recalling-civil-relationship-with-segregationists. 13 September 2021.
29. "Spokesvultures for Ecological Awareness: an interview with Timothy Morton." go.gale.com/ps/i.do?id =

GALE%7CA496566676&sid = googleScholar&v = 2.1&it = r&linkaccess = abs&issn = 03,170,802&p = AONE&sw = w&userGroupName = nysl_oweb&isGeoAuthType = true. 14 October 2021.

30. Ibid.
30. Mosley. *NY Times*. www.nytimes.com/2019/09/06/opinion/sunday/walter-mosley.html. 6 September 2019.

Social Syntax

1. *Descent of Man*. Global Grey eBooks, 2018. Based on Second edition 1874. 125.
2. *The Politics*. Carnes Lord. University of Chicago Press, 2013. Book 1.2.
3. *Ibid*.
4. Brian Goodwin. *How the Leopard Changed his Spots*. Scribner, 1994. 320p.
5. Robert Boyd & Peter Richerson. Philos *Trans R Soc Lond B Biol Sci*. 2009 November 12; 364(1533): 3281–3288. https://doi.org/10.1098/rstb.2009.0134.
6. Sung-Ha Hwang & Samuel Bowles. *Journal of Economic Behavior & Organization*, Vol. 83, No. 3, August 2012. 330–341. https://doi.org/10.1016/j.jebo.2012.06.001.
7. Rachel Croson. "Theories of Commitment, Altruism and Reciprocity: Evidence from Linear Public Goods Games." *Economic Inquiry*, Vol. 45, No. 2, 2007. 199–216 https://doi.org/10.1111/j.1465-7295.2006.00006.x.
8. Nancy Folbre. "Evolution and Intersectional Conflict." https://blogs.umass.edu/folbre/2021/01/18/evolution-and-intersectional-conflict/. 20 June 2021.
9. Ballotopedia: http://ballotpedia.org/Wisconsin_collective_bargaining. 30 June 2021.
10. Sung-Ha Hwang. *Op. Cit*.
11. Environmental Defense Fund. www.edf.org/sites/default/files/Business-and-the-Fourth-Wave-of-Environmentalism_2019.pdf. 15 March 2020.

12. Seymour Burchman. *Harvard Business Review*. https://hbr.org/2018/11/how-to-tie-executive-compensation-to-sustainability. 12 December 2018.
13. Wilson. *Op. Cit.* 78.
14. Lewis Carroll. *Through the Looking Glass*. Createspace, 2017. Chapter II.
15. Michael A. Brockhurst. "Sex, Death, and the Red Queen." *Science*. July 8, 2011.
16. Holly Melgard. Roof Books, 2021. 108p.
17. Clark Coolidge. "Trilobite." *Space*. Harper and Row, 1970. 68.
18. *The Drawdown Review*. Project Drawdown, 2020. 5.
19. Hank Lazer. *Opposing Poetries*. Northwestern University Press, 1996. 82.
20. Hélène Aji. "'Writing (as) (and) thinking': Charles Bernstein's Work 'in' Language." *Études anglaises* 2006/3 (Vol. 59). 341.
21. *Red Shift*. Roof, 1988.
22. Ron Silliman. "Realism: An Anthology of Language Writing." *Ironwood*. Ironwood Press, 1982. 68.
23. $L = A = N = G = U = A = G = E$, 1. February, 1978.
24. Gary Snyder. "Rip Rap." *Rip Rap and Cold Mountain Poems*. Shoemaker & Hoard, 2003.
25. *Down to Earth*. Op. Cit. 13.
26. Che Qianzi. "Magnificent." *Urban Poetry from China*. Op. Cit. 48.
27. Ferdinand Braudel. *Civilization and Capitalism, 15th–18th Century*. Harper & Row, 1984. Vol. 2, 455.
28. Latour. *Op.Cit.* 21.
29. quoted in John Hale. "Corporations Are Changing. Exhibit No. 1: Larry Fink's Letter." www.morningstar.com/articles/1075393/corporations-are-changing-exhibit-no-1-larry-finks-letter. 22 January 2022.
30. Latour. *Op. Cit.* 23.
31. Hugh Patterson. "The Competitive Darwin." *Paleobiology*, 31(2), 2005. 56. https://doi.org/10.1666/0094-8373(2005)031[0056:TCD]2.0.CO;2.
32. Michael Bulmer. "The Theory of Natural Selection of Alfred Russel Wallace FRS." *Notes and Records of the Royal Society*. 2005, 59. 125–136.

33. Bob Perelman. "To My Mother." *The Future of Memory*. Roof, 1998. 82.

Notes: How Can Culture Change Habitat?

1. Based on author's meetings with 350.org management after the Occupy Wall Street protests in September, 2011.
2. "Deep Adaptation: A Map for Navigating Climate Tragedy" Revised 2nd Edition July 27, 2020. IFLAS Occasional Papers Volume 2. University of Cumbria, Ambleside, UK. http://insight.cumbria.ac.uk/id/eprint/4166/.
3. Moore. Op. Cit.
4. Cecilia Capuzzi Simon. *NY Times*. www.nytimes.com/2015/04/12/education/edlife/12edl-12mfa.html 9 April 2015.
5. "Hacking…" www.youtube.com/watch?v=trxzyEfG8cY. Black Hat. 3 August 2019.
6. *Ibid.*
7. Armey Curve. I have simplified the curve to show only basic information from: thinkbynumbers.org/economics/gdp/ideal-level-of-government-spending/. 20 June 2015. And analysis of Armey Curve at: www.researchgate.net/profile/Cem-Isik/publication/352867278_The_Armey_Curve_Hypothesis-based_Environmental_Kuznets_Curve_EKC_Hypothesis_Testing_Across_the_US_States_with_Government_Spending/links/61692ffe951b3574c63ff683/The-Armey-Curve-Hypothesis-based-Environmental-Kuznets-Curve-EKC-Hypothesis-Testing-Across-the-US-States-with-Government-Spending.pdf.
8. www.cato.org/publications/commentary/optimum-government, 20 January 2009.
9. "Sunday Morning." *The Collected Poems*. Vintage, 2015.
10. http://webhome.cs.uvic.ca/~mserra/AttachedFiles/PersonalPolitical.pdf. 13 February 2022.
11. "Bubble Wrap." *Money Shot*. Wesleyan University Press, 2011.
12. "Hello the Roses." *Hello the Roses*. Op. Cit.
13. *Black and Blur (consent not to be a single being)*. Duke, 2017. Vii.
14. "Song Of." *Echolocation*. Roof, 2018. 19.

Groups Build Social Syntax

1. Jorge Luis Borges, "The Nothingness of Personality" (1922). *Selected Non-Fictions*. Viking, 1999. 4.
2. "Preface." *Leaves of Grass*. American Renaissance, 2009. ix.
3. "For the Sake of People's Poetry: Walt Whitman and the Rest of Us." *Passion: New Poems 1977–1980, with a preface by the poet*. Beacon Press, 1980. 252. www.poetryfoundation.org/articles/68627/for-the-sake-of-peoples-poetry. 27 October 2021.
4. Moore. Op. Cit. 8, 9.
5. *Ibid*. 19.
6. Wilson. *Op. Cit*. 19.
7. Hyunju Kim et al. "Universal Scaling Across Biochemical Networks on Earth." *Science Advances*. 16 January 2019: Vol. 5, No. 1, eaau0149. Abstract: https://doi.org/10.1126/sciadv.aau0149.
8. Keller, Lynn. "Twenty-First-Century Ecopoetry and the Scalar Challenges of the Anthropocene." *The News from Poems: Essays on the 21st-Century American Poetry of Engagement*, ed. Jeffrey Gray and Ann Keniston, University of Michigan Press, 2016, 51. www.jstor.org/stable/10.3998/mpub.9227505.6.
9. *Ibid*. 52.
10. Vernadsky. *Op. Cit*., 1998. 25.
11. Derek Woods. "Scale Critique for the Anthropocene." *Minnesota Review*, Issue 83, 2014 (New Series). 134.
12. Keller in Gray & Keniston. *Op. Cit*. 53.
13. "Terminal Poetics of the Anthropocene." *The Cambridge Companion to 21st Century American Poetry*, ed. Timothy Yu. Cambridge University Press, 2021. 163.
14. "Scale." *Telemorphosis: Theory in the Era of Climate Change*, vol. 1, ed. Tom Cohen Open Humanities Press, 2012. 148.
15. Bernat Corominas-Murtra. "A Unified Approach to the Emergence of Complex Communication." Universitat Pompeu Fabra. http://digital.csic.es/bitstream/10261/46559/1/tbc.pdf. 13 February 2022.
16. Luc Steels. "Modeling the Cultural Evolution of Language" *Physics of Life Reviews* 8 (2011). 339–356.
17. Monica A. Gorzelak et al. "Inter-Plant Communication Through Mycorrhizal Networks Mediates Complex Adaptive Behaviour in Plant Communities." *AoB Plants*. 2015; 7: plv050.

18. Humboldt. Sagwan Press, 2018. 438p.
19. Vaclav Smil. *The Earth's Biosphere*. MIT Press, 2003. 256p.

Some Structures of Syntax—Parataxis & Hypotaxis

1. Bob Perelman. *The Marginalization of Poetry: Language Writing and Literary History*. Princeton University Press, 1996. 60.
2. Trish Salah. Roof Books, 2014. 11.
3. *The Age of Huts*. University of California Press, 2007. 101.
4. "Variations: A Return of Words." *A Thought is the Bride of What Thinking*. Tuumba, 1976. unpaginated
5. *Citizen*. Greywolf Press, 2014. 139–146.
6. Roof Books, 2018. 201p.
7. Che Qianzi. "A Song," tr. Jeff Twitchell. jacketmagazine.com/20/pt-chinese.html. 15 November 2019.
8. James Sherry. "Snowball Earth Creates Innovative Poetry." *Oops!: Environmental Poetics*. BlazeVox, 2013. 133–142.
9. *The Undercommons: Fugitive Planning & Black Study*. Autonomedia/Minor Compositions, 2013. 108p.
10. A list of books using the metaphor that can be construed as material. www.google.com/search?tbm=bks&hl=en&q=value+corporate+takeovers+ecosystem. 25 June 2018.
11. *Girls on the Run*. FSG, 1999. 8.
12. I discuss this issue in the essay "Against One Model Alone." *Aufgabe* 13, 2014.

Scalable Syntax: Poetry Model of the Biosphere

1. Hyunju Kim et al. "Universal Scaling Across Biochemical Networks on Earth." *Science Advances*. 16 January 2019: Vol. 5, No. 1, eaau0149. "Introduction." https://doi.org/10.1126/sciadv.aau0149.
2. Ibid.
3. Michael T. Hannan and John Freedman. "The Population Ecology of Organizations." *The American Journal of Sociology*, Vol. 82, No. 5. March 1977. 929–964 cited on "Environment and

Ecology" website. https://environment-ecology.com/history-of-ecology/132-history-of-ecology.html. 13 February 2022.

4. Carolyn Merchant. *The Death of Nature: Women, Ecology, and the Scientific Revolution*. HarperOne, 1990. 69.
5. Kim et al. *Op. Cit.* 1.
6. "A Defense of Poetry." *A Defense of Poetry and Other Essays*. Kindle Edition, 2004. Paragraph 1.
7. New Directions, 2016. 84p.
8. E. Summerson Carr & Michael Lempert. *Scale: Discourse and Dimensions of Social Life*. University of California Press, 2016. 14.
9. Note on scaling, form, and behavior: Evading, avoiding, and eliminating form and syntactical order, that is, the effects of the biosphere on poetry and translation, has not promoted more ecologically sound solutions any more than how freedom of invention in the modern age of industry has polluted the planet. Contemporary young people's social solutions, initiated by assertion of self and more clearly defined limits on behavior than in my post-World War II generation, mostly reconfigure received components—controls, forms, limits, sequences, words, and phrases in their language and others. It's the order of behavior that counts as much as what those behaviors are: "Every power relation is not bad in itself, but it is a fact that always involves danger" (Michel Foucault. *Politics, Philosophy, Culture*. Routledge, 1990. 168).
10. Note: This imbalance of critical attention and distribution methods is relative. Even with increasing numbers of reviews and essays, the same few names appear in review organs while there are many new books each year worth serious attention.
11. See chapter on external cognition "Snowball Earth Creates Innovative Poetry" in *Oops! Environmental Poetics* BlazeVox, 2013. 133–142. Specific reference to external cognition: Mark Rowlands. "Environmental Epistemology." *Ethics & the Environment*, Vol. 10, No. 2, 2005.
12. "The Interminable Uncanniness." *Psychoanalysis at the political border*. Ed. Leo Rangell and Rena Moses-Hrushovski. International Universities Press, 1996. Chapter 7.

Notes: The Anthropocene: Ecosystems & Time Frames

1. *Plows, Plagues, and Petroleum: How Humans Took Control of Climate*. Princeton, 2016. 240p.
2. "Defining the Anthropocene." *Nature* 519. March 12, 2015. 171–180.
3. Eduardo Viveiros de Castro & Deborah Danowski. "Humans and Terrans in the Gaia War." *A World of Many Worlds*. Duke, 2018. 176. https://doi.org/10.1215/9781478004318-007.
4. Moore. *Op. Cit.* 25.
5. Quoted in Derek Woods. "Scale Critique for the Anthropocene." *Minnesota Review*. Duke University Press, 2014. Issue 83, 133.
6. Woods. *Ibid.*
7. Woods. *Op. Cit.* 134.
8. This list is far too sketchy. I have expanded on the questions around several contemporary states' governance in *The Oligarch*. (Palgrave, 2018. 158p). For readers interested in nation states and pollution, many histories offer nuanced details on how state and regional cultures have absorbed demands of petroleum, agriculture, and power production in growing populations. Works of Michael J. Watts on the Nigerian Petrostate, Meine Van on Indonesian agriculture, Dean Spears on pollution in India, and Sergio Margulis on Brazilian deforestation. The literature is vast and beyond the scope of this book.
9. Von Uexküll. *Op. Cit.* 133.
10. *Ibid.* 135.
11. Roy Morgan. www.roymorgan.com/findings/8044-most-important-problems-facing-australia-world-july-2019-201907020950. 11 August 2019.
12. *Ecocriticism: Environments in Anglophone Literatures*. Eds. Sonja Frenzel & Birgit Neumann. Anglistik and englischunterricht, 2017. 169.
13. Tr. John Bednarz. University of Chicago, 1989. 200p.
14. Vernadsky. *Op. Cit.* 52.
15. Ibid. 67.
16. Deleuze & Guattari. *What is Philosophy*? Verso, 2003. 101.

Networks of Metaphor

1. David Harvey "The Nature of Environment: The Dialectics of Social and Environmental Changes." *The Socialist Register*. https://socialistregister.com. 15
2. Donald Schön. "Generative Metaphor: A Perspective on Problem-Setting in Social Policy." Ortony. *Op. Cit.* 138
3. Mark Erickson. "Network as Metaphor." *International Journal of Criminology and Sociological Theory.* Vol. 5, No. 2 (2012). 912. https://ijcst.journals.yorku.ca/index.php/ijcst/article/view/35701. Accessed 12 June 2022.
4. W. Baldamus "Networks." *The Sociology of Wilhelm Baldamus: Paradox and Inference*. Eds. M. Erickson, M. and C. Turner, C. Ashgate Publishing, 2010 (1982). 107.
5. "The Conduit Metaphor: A Case of Frame Conflict in Our Language about Language," *Metaphor and Thought*. Ed. Andrew Ortony. Cambridge University Press, 1993. 168.
6. *The Language of Metaphors*. Routledge, 2011. 2.
7. Jacques Lacan. *Ecrits*. W. W. Norton, 2002. 16.
8. The Five Graces Group. "Language is a Complex Adaptive System." Language Research Learning Club, University of Michigan, 2009. 5ff.
9. George Lakoff & Mark Johnson. *Metaphors We Live By*. University of Chicago Press, 1980. 256p.
10. *The Unfollowing 77.* Omnidawn, 2016. 89.
11. Tenenbaum. et al., "How to Grow a Mind: Statistics, Structure, and Abstraction." *Science*, 11 Mar 2011: Vol. 331, Issue 6022, 1279–1285.
12. *Post Classic*. Roof, 2019. 9.
13. Ortony. *Op. Cit.* 202–252.
14. Reddy. *Op. Cit.* 167.
15. Erving Goffman's frame analysis is easily adapted to environmentalism. *Frame analysis: An essay on the organization of experience*. Harper and Row, 1974. 586p.
16. Reddy. *Op. Cit.* 165.
17. Distinguishing between problematic categories that increase the need for further investigation and imaginary, fabricated, and intangible categories leads us to look at things with their surroundings.
18. Reddy. *Op. Cit.* 165.
19. Graziano. *Consciousness*. *Op. Cit.* 32.

20. Discussed in *Oops!* Op. Cit. 137ff.
21. Lakoff. *Op. Cit.* 218,19.
22. *Galileo's Commandment*. Ed. Edmund Blair Bolles. W.H. Freeman, 1997. 310.
23. Yossi Arzouan, et al. "Dynamics of Hemispheric Activity During Metaphor Comprehension: Electrophysiological Measures." https://doi.org/10.1016/j.neuroimage. 15 February 2007.
24. Kathleen Ahrens et al. "Functional MRI of conventional and anomalous metaphors in Mandarin Chinese." https://doi.org/10.1016/j.bandl.2005.10.004.
25. A series of references supporting more detail than I can process here is found at: https://link.springer.com/chapter/10.1007/978%E2%80%930-387%E2%80%9374,980-8_8. 10 August 2019.
26. Lakoff in Otony. *Op. Cit.* 229.
27. Uexküll. *Op. Cit.* 129–130.
28. Champika Epa Ranasinghe & Christian Kray. www.researchgate.net/figure/Signal-propagation-reflection-refraction-and-diffraction_fig3_329009261. 13 February 2022.
29. *Entanglements*. Wesleyan, 2017.

Environmental Identities

1. Daniel P. Todes. *Darwin without Malthus: The Struggle for Existence in Russian Evolutionary Thought*. Oxford, 1989. 104.
2. *Ibid.* 170.
3. Giles Deleuze. *Difference and Repetition*. Tr. Paul Patton. Columbia, 1994. 170.
4. *Ibid.* 168.
5. *Ibid.* 169.
6. Barad. *Op. Cit.*
7. Talcott Parsons (1958) "Social Structure and the Development of Personality." *Psychiatry*, Vol. 21, No. 4. 321–340, https://doi.org/10.1080/00332747.1958.11023143.
8. Almut Höfert. *Describing the Enemy*. Campus Historiche Studien, 2003. 76.
9. *The Language Letters*. Op. Cit.
10. Gertrude Stein. *How to Write*. Dover, 1975. 25.
11. Darwin, Gould, Wilson et al. all write about the functions of nature rather than efficiency or perfection.

12. *Encyclopedia of Global Environmental Change. Vol. 5: Social and Economic Dimensions of Global Environmental Change*, ed Peter Timmerman. John Wiley & Sons Inc., 2002. 384–398.
13. Clark Coolidge. *The Book of During*. The Figures, 1991. 112.
14. *Deep Adaptation: A Map for Navigating Climate Tragedy Revised*. Eds. Jem Bendell & Rupert Read. 2nd edition released July 27th, 2020 and several versions: https://jembendell.com/2019/05/15/deep-adaptation-versions/. 13 February 2022.

Environmental Autobiography

1. Thomas Fink. dichtungyammer.wordpress.com/. 12 December 2021.
2. Erkki Patokorpi. "Information Pluralism and Some Informative Modes of Ignorance." *Information*. 2011, 2, 50. https://doi.org/10.3390/info2010041.
3. Afsaneh Najmabadi. *Professing Selves*. Duke University Press, 2014. Introduction.
4. "Scripture." *Just Saying*. Wesleyan, 2013.
5. "Xenophobia." *Extremities*. The Figures, 1978.

Social Syntax II: Linkages & Connectors

1. Raj Chetty, et al., "Where Is the Land of Opportunity? The Geography of Intergenerational Mobility in the United States." *The Quarterly Journal of Economics*. June, 2014. https://academic.oup.com/qje/article/129/4/1553/1853754. 20 October 2021.
2. Piketty, *Op. Cit.*
3. *The Grand Piano*. Mode A. This Press, 2007 and *The Language Letters. Op. Cit.*
4. *The Language Letters. Op. Cit.*
5. "On Speech," *This* 1. 1971.
6. www.writing.upenn.edu/~afilreis/88/langpo.html. 2. www.encyclopedia.com/arts/educational-magazines/paradoxes-and-oxymorons. 3. www.modernamericanpoetry.org/content/michael-miller-john-ashbery. 5 July 2019.

7. Justin Gillis, "Flooding of Coast, Caused by Global Warming, Has Already Begun." 3 September 2016.

Identity's Constructive & Connective Ecosystem

1. Lucie Salwiczek, et.al. "What Do Animals Remember About Their Past?" *Learning and Memory: A Comprehensive Reference.* 1. 441–459. 2008. https://doi.org/10.1016/B978-012370509-9.00060-7.
2. Graziano. *Consciousness. Op. Cit.* Frontispiece.
3. *Ibid.* 13.
4. *Ibid.* 19.
5. Image adapted from Graziano. *Ibid.* 18.
6. *Ibid.* 19.
7. Tenenbaum, Op. Cit.
8. Reddy. *Op. Cit.* 169.
9. *Ibid.* 171, 2.
10. *Ibid.* 172.
11. Discussed in Scott Pound. "The Chinese Notebook and the Materialities of Communication." vectors.usc.edu/thoughtmesh/publish/191.php#. 7 July 2018.
12. *Maximus*, III 37. University of California Press, 1985. 664p.
13. Image built from Graziano's prose. *Consciousness. Op. Cit.* 125–6.
14. Premack. *Op. Cit.*
15. *Ibid.* 159–162.
16. Marshall McLuhan. *Understanding Media.* MIT Press, 1994. 8.
17. *On Intersectionality: Essential Writings.* The New Press, 2022. 320p.
18. *Feminist Theory: from margin to center*. Routledge, 1984. 14.
19. Maurice Bloch. *Going In and Out of Each Other's Bodies*. Routledge, 2012. 172p.
20. Maurice Bloch. "The resurrection of the house amongst the Zafimaniry of Madagascar." *About the House: Lévi-Strauss and Beyond*, eds. J. Carsten & S. Hugh-Jones. Cambridge University Press, 1995. 77, 8.
21. ich.unesco.org/en/RL/woodcrafting-knowledge-of-the-zafimaniry-00080. 13 July 2019.
22. time.com/5669061/china-climate-change/ 7 July 2020.

23. These discourses within our primary exchanges show the limitations of a purely ecological approach to climate. The performances of individual identities coalesce through poorly defined, composite channels. But has the information model swallowed critical theory whole? Is there any non-communications space in which to critique information models of the biosphere? The dialectic predicts that the critique of information models appears in information. But since information society operates by evaluation like academic discourse, a dialectical critique of climate modeling seems easy to present in similar terms. Inside/outside and network/field ideas must be included for the ecological model to be absorbed into society that shows concern for telos, eschatology, and causation. Scott Lash's information model agrees with environmentalism in looking less at causality than its constructive process that excludes classes exploited in Marxism. The model of exclusion rhymes better with Bruno Latour's suggestion (*Down to Earth*. Op. Cit.) that leading classes as I have presented in *The Oligarch* exclude the masses from climate solutions reducing them to survivalist regimes and exposing them to autocracy as populations peak and fall later this century. This dual performance of climate change shows how the partite construction of our world renders the inside/outside model less serviceable than in the Hegelian. The performative method of building identity describes network interactions quite differently from a communications model (Scott Lash. *Critique of Information*. Sage, 2002. 234p).
24. Roof, 2014. 5.
25. Verso, 2010. 336p.
26. "Conceptualizing the Field." jacket2.org/commentary/jonathan-skinner. 8 July 2020.
27. *This Magazine*. No. 7 (Spring 1976), ed. Barrett Watten. 3.
28. *Textual Politics and the Language Poets*. Indiana University Press, 1989. 6.
29. *Roof Magazine*. #4: Washington, DC. #5: San Francisco. Ed. Sherry et al. eclipsearchive.org/projects/ROOF/roof.html. 20 October 2021.
30. Reuters. "Saudi Arabia sees domestic energy use falling, plans renewables push." www.reuters.com/article/us-saudi-energy-reforms/saudi-arabia-sees-domestic-energy-use-falling-plans-renewables-push-idUSKCN1P918N. 15 January 2019.

31. Common theories of the organization of organisms are found summarized here: https://plato.stanford.edu/entries/levels-org-biology/. 2 March 2022.
32. Cathy Cohen. signsjournal.org/ask-a-feminist-cohen-jackson/, 17 March 2020.
33. Seven Stories Press, 2004. 491p.
34. Crenshaw. *Op. Cit.*
35. Donna Haraway, *When Species Meet*. University of Minnesota Press, 2008. 301 & 70–71.
36. George Herbert Mead. *The Philosophy of Act*. University of Chicago Press, 1938. 115

The Condition & Hierarchy of Identification

1. Penguin, 1982. 208p.
2. Fence, 2016. 80p.
3. "Mutability." *John Keats and Percy Bysshe Shelley Complete Poetical Works*. The Modern Library, 1940. 562.
4. Princeton University Press, 2014. 432p.
5. *The Structure of Evolutionary Theory*. Belnap, 2002. 922.
6. Fiona Templeton. *Cells of Release*. Roof, 1997. Cell 4.
7. Roof, 2011. 84p.
8. BlazeVox, 2015. 148p.

Impact of Text & Environment on Another Self

1. "Poetry Wars, 1984—Documents from *Poetry Flash*, including 'Of Duncan and Zukofsky on Film: Traces Now and Then' by David Levi Strauss." *Dispatches from the Poetry Wars*, eds. Michael Boughn & Kent Johnson. Posted in Simon Fraser University archived websites, 2020. https://wayback.archive-it.org/. 13 February 2022. www.dispatchespoetrywars.com/documents/poetry-wars-1978-documents-from-the-poetry-flash-including-of-dunncan-and-zukofsky-on-film-traces-now-and-then-by-david-levi-strauss/. 12 February 2022.

2. Juan Manuel Rodríguez Caso. Universidad Autonoma de Mexico. www.academia.edu/1675526/Darwin_and_Wallace_on_Morals_and_Ethics_Two_Different_Views_from_Natural_Selection. 13 February 2022.
3. Claude Levi- Strauss. "Translators Preface." *Structural Anthropology*. Basic Books, 2008. 432p.
4. *Ibid.*
5. Both kin and groups shown to be mathematically similar: Jos Kramer and Joël Meunier. www.ncbi.nlm.nih.gov/pmc/articles/PMC4850877/. 28 September 2019.
6. Omar Tonsi Eldakar & David Sloan Wilson. "Selfishness as Second-order Altruism." Proc Natl Acad Sci U S A. May 13, 2008; 105(19): 6982–6986. www.ncbi.nlm.nih.gov/pmc/articles/PMC2383986/. 12 February 2022.
7. Alexie Sherman. www.poemhunter.com/poem/evolution-4/. 12 February 2022.
8. James Clerk Maxwell. www.poetryfoundation.org/poems/45777/molecular-evolution. 12 February 2022.

Works Cited

Introduction

Alexander, Will. *The Combustion Cycle*. Roof Books, 2021.

Auden, W. H. "In Memory of W.B. Yeats." *Collected Poems*, ed. Edward Mendelson. Vintage, 1991.

Australian Academy of Sciences. www.science.org.au/learning/general-audience/science-climate-change/2-how-has-climate-changed. 12/29/2021.

Baudelaire, Charles. "Edgar Allan Poe: His Life and Works." *The Works of Edgar Allan Poe*, trans. H. Curwen. John Camden Hotten, 1873. www.eapoe.org/papers/misc1851/1873000m.htm. 5 February 2022.

Bergvall, Caroline. *Drift*, Nightboat Books, 2014.

Bernays, Edward. *The Engineering of Consent*. University of Oklahoma Press, 1955.

Brooks, Gwendolyn. "Poetry Is Life Distilled" quoted from Brooks and the title of her biography by Christine M. Hill. Enslow Publishing, 2008.

Dickinson, Emily. "This Was a Poet—It Is That." *The Poems of Emily Dickinson*. #446, ed. R.W. Franklin. Belnap, 1999.

Gould, Stephen Jay. *Time's Arrow, Time's Cycle*. Harvard University Press, 1988.

Grenier, Bob. "Northern New Hampshire." *This Magazine*, #7. This Press, 1976.

Hartley, George. "Textual Politics and the Language Poets." www.writing.upenn.edu/~afilreis/88/hartley.html. 22 November 2020.

Hejinian, Lyn. "The Rejection of Closure." *The Language of Inquiry*. University of California Press, 2000.

Hill, Alice and Martinez-Diaz, Leonardo. "Adapt or Perish." *Foreign Affairs*. January/February 2020.

J. Sherry, *Selfie*, https://doi.org/10.1007/978-981-19-4870-1

hooks, bell. *Wounds of Passion: A Writing Life*. Holt, 1999. 288p.
McIntosh, Emma. "No, a Republican Strategist Didn't Invent the Term 'Climate Change." *Canada's National Observer*. www.nationalobserver.com/2019/08/01/news/no-republican-strategist-didnt-invent-term-climate-change. 6 January 2022.
Perch-Nielsen, Sabine L., Bättig, Michèle B., and Imboden, Dieter. "Exploring the Link Between Climate Change and Migration." *Climatic Change* 91(3–4). https://doi.org/10.3929/ethz-b-000014553. 7 February 2022.
Plato. *The Republic*. Book VII., trans. Allan Bloom. Basic Books, 1968.
Pound, Ezra. *ABC of Reading*. Faber, 1934.
Rasula, Jed. *This Compost*. University of Georgia Press, 2012.
Report on the Impact of Climate Change on Migration. Joe Biden's White House Report. October, 2021. reliefweb.int/report/world/report-impact-climate-change-migration-october-2021. 22 November 2021.
"Selfie." *OED*. 2016.
Shoptaw, John. "Why Ecopoetry?" www.poetryfoundation.org/poetrymagazine/articles/70299/why-ecopoetry. 12 December 2021.
Stein, Gertrude. "Poetry and Grammar." *Lectures in America*. Beacon Press, 1957.
Stevens, Wallace. "Man Carrying Thing." *The Collected Poems of Wallace Stevens*. Knopf, 1954.
Thoreau, Henry David. "Nature." poets.org/poem/nature-1. 5 Feburary 2022.
"U.S. Global Change research Program." *The Climate Report*. Melville House, 2019.
Valéry, Paul. "Poetry and Abstract Thought," trans. Charles Guenther. *The Kenyon Review,* Spring, 1954.
Vicuña, Cecilia. "Performances." www.ceciliavicuna.com/performances. 26 December 2021.
Von Dawans, Bernadette. *PLoS One*. 2018; 13(10): e0204665. www.ncbi.nlm.nih.gov/pmc/articles/PMC6201881/. 8 January 2022.
Wheeler, Raymond Holder. "The Effect of Climate on Human Behavior in History." *Transactions of the Kansas Academy of Science*. Vol. 4, Apr. 10, 1943.
Wordsworth, William. "Preface." *Lyrical Ballads*. Bartleby.com/39. 28 December 2021.

The Particles

Barabasi, Albert-Laszlo. *The New Science of Networks*. Perseus, 2002.
Bernstein, Charles. "Poetic Citizenship and Negative Dialectics." *topsy-turvy*. University of Chicago Press, 2020.

Bezerra, Erickinson. "The Meaning(s) of Declassification in Antonio García Gutiérrez." www.academia.edu/50231535/The_meaning_s_of_the_declassification_in_Antonio_Garc%C3%ADa_Guti%C3%A9rrez?email_work_card=thumbnail. 12 October 2021.

Bromige, David. *My Poetry.* The Figures, 1980.

D'Argembeau, Arnaud. "On the Role of the Ventromedial Prefrontal Cortex in Self-Processing: The Valuation Hypothesis." *Frontiers in Human Neuroscience*, 2013. https://doi.org/10.3389/fnhum.2013.00372.

Darwin, Charles. *The Origin of Species*. Modern Library, 1998.

de Beauvoir, Simone. *The Second Sex.* Jonathan Cape, 1956.

Foertsch, Julie and Gernsbacher, Morton Ann. "IN SEARCH OF GENDER NEUTRALITY: Is Singular *They* a Cognitively Efficient Substitute for Generic *He*?." Psychol Sci. 1997 March; 8(2): 106–111. https://doi.org/10.1111/j.1467-9280.1997.tb00691.x, https://doi.org/10.1111/j.1467-9280.1997.tb00691.x. 7 February 2022.

Gardner, Drew, Gordon, Nada, et.al., eds. *Flarf: An Anthology of Flarf*. Edge Books, 2017.

Ghazanfar, Asif A. and Rendall, Drew. "Evolution of Human Vocal Production." *Current Biology*. Vol. 18, No. 11. R457. www.cell.com/current-biology/pdf/S0960-9822(08)00371-0.pdf. 7 February 2022.

Gould, Stephen Jay. *The Structure of Evolutionary Theory*. Belnap, 2002.

Gutierrez, Antonio Garcia. *Ojos de la arena. ejercicios de desclasificación*, trans. Erickinson Bezerra. ACCI Ediciones, 2020.

Hayden, Robert. *Phylon*, VI, vol. 3. Clark Atlantic University, 1945.

Liberman, Zoe, et al.. "The Origins of Social Categorization." *Trends in Cognitive Sciences*, July, 2017. https://doi.org/10.1016/j.tics.2017.04.004.

Pievani, Telmo. *Imperfezione*. Raffaello Cortina Editore, 2019.

Pope, Alexander. *An Essay on Man*, ed. Tom Jones. Princeton University Press, 2016.

Premack, David and Woodruff, Guy. "Does the Chimpanzee Have a Theory of Mind?". *The Behavioral and Brain Sciences*, 1978. Published online by Cambridge University Press, 2010. www.cambridge.org/core/journals/behavioral-and-brain-sciences/article/does-the-chimpanzee-have-a-theory-of-mind/1E96B02CD9850016B7C93BC6D2FEF1D0.

Reilly, Evelyn. "Self." *Echolocation*. Roof, 2018.

Rothenberg, David. *Survival of the Beautiful*, Bloomsbury Press, 2011.

Smart, Christopher. "Jubilate Agno." *The Poetical Works of Christopher Smart, I: Jubilate Agno*, ed. Karina Williamson. Clarendon, 1980.

Summers, Richard L. "The Imperative of Self-Reference in a Theoretical Framework for Biology." www.academia.edu/50664004/The_imperative_of_self_reference_in_a_theoretical_framework_for_biology. 11 October 2021.

Tardos, Anne. *Both Poems*. Roof Books, 2011.
Thanukos, Anastasia. "Darwin's 'Extreme' Imperfection?" *Evo Edu Outreach* 2. https://doi.org/10.1007/s12052-008-0105-0. 2009.

Nouns & Things: Changing Their Climate

Armantrout, Rae. *Conjure*. Wesleyan, 2020.
Barad, Karen. *Meeting the Universe Halfway*. Duke University Press, 2007.
———. "Posthumanist Performativity: Toward an Understanding of How Matter Comes to Matter." *Signs: Journal of Women in Culture and Society*, vol. 28, no. 3, Spring 2003. University of Chicago Press, 2003. 802.
Chen, Weile, et al. "Root Morphology and Mycorrhizal Symbioses Together Shape Nutrient Foraging Strategies of Temperate Trees." PNAS August 2, 2016 113 (31) 8741–8746. https://doi.org/10.1073/pnas.1601006113.
Godfrey-Smith, Peter. *Other Minds*. FSG, 2016.
Graziano, Michael S. *Consciousness and the Social Brain*. Oxford University Press, 2013.
Harvey, David. *The Nature of Environment: The Dialectics of Social and Environmental Change*. Presented at Havens Center of the University of Wisconsin, 1991. http://socialistregister.com/index.php/srv/article/view/5621/2519. 15 February 2022.
Kant, Immanuel. *Critique of Pure Reason*. Cambridge University Press, 1999.
Latour, Bruno. *Reassembling the Social*. Oxford University Press, 2007.
Latour, Bruno. "On actor-network theory: A few clarifications." *Soziale Welt, 47. Jahrg., H. 4*, 1996. www.jstor.org/stable/40878163?origin=JSTOR-pdf. 16 September 2020.
Latour, Bruno and Woolgar, Steve. *Science in Action*. Harvard, 1987.
Moore, Jason. *Capitalism and the Web of Life*. Verso, 2015.
Morton, Timothy. *Ecology without Nature*. Harvard, 2007.
Olson, Charles. "Projective Verse." *Collected Prose*, eds. Don Allen and Benjamin Friedlander. University of California Press, 1997.
Pollan, Michael. *Botany of Desire*. Random House, 2002.
Robinson, Mary. "Sonnet to Evening: Written Under a Tree in the Woods of St. Amand, in Flanders." *Poems*. J. Bell, 1791.
Stein, Gertrude. *Lectures in America*. Beacon Press, 1957.
Vernadsky, Vladimir. *The Biosphere*. Copernicus, 1998.
Waldman, Anne. *Fast Speaking Woman and other Chants*. City Lights, 1975.
Whitman, Walt. *Leaves of Grass*. Library of America, 2018.

Verb Solutions Adapt

Ariely, Dan. *Predictably Irrational*. HarperCollins, 2009.
Barad, Karen. *Meeting the Universe Halfway*. Duke University Press, 2007.
Fischer, Norman. *Museum of Capitalism*. Talisman House, 2021.
Godfrey-Smith, Peter. *Other Minds*. FSG, 2016.
Inman, Peter. "6 + 5 pieces from Webern (2015, 2017)." *Scratches*. if p then q, 2019.
Kahneman, Daniel, et al., eds. *Judgment under uncertainty: Heuristics and biases*. Cambridge University Press, 1982.
Milton, John. *Paradise Lost*. Book V, 1667. Penguin Classics, 2012.
Moore, Jason. *Capitalism and the Web of Life*. Verso, 2015.
Niedecker, Lorine. *Collected Works*, ed. Jenny Penberthy. University of California Press, 2002.
O'Hara, Frank. "Second Avenue." *The Collected Poems of Frank O'Hara*, ed. Donald Allen. University of California Press, 1995.
Perelman, Bob. "Chronic Meanings." *Virtual Reality*. Roof, 2019.
Sherry, James. "Part 1: Business Process Interruption and Risk Homeostasis: Analysis for Technology Solutions to Regulatory Impact." *Oops! Environmental Poetics*. BlazeVox, 2013.
Stein, Gertrude. "Poetry and Grammar." *Lectures in America*. Beacon Press, 1957.
Sunstein, Cass. *Nudge*. Yale University Press, 2008.
The Climate Report. U.S. Global Change Research Program. Melville House, 2018.
The Nature Conservancy. "Let's Talk Climate Change: A How-to Guide." www.nature.org/en-us/what-we-do/our-priorities/tackle-climate-change/climate-change-stories/can-we-talk-climate/. 12 December 2019.
Wilson, David Sloan. *Darwin's Cathedral*. University of Chicago Press, 2002.

Borderlands: Am I I & They? Knowing Without Or

Anzaldúa, Gloria. *La Frontera/Borderlands*. Aunt Lute Books, 1999. 288p.
Baraka, Amiri. "Short Speech to My Friends." *African American Poetry*, ed. Kevin Young. The Library of America, 2020.
Buck, Marie. *Goodnight, Marie, May God Have Mercy on Your Soul*. Roof, 2017.
Cure, The. "This. Here and Now. With You." www.thecure.com/lyrics/this-here-and-now-with-you/?msclkid=4991. 2 February 2022.
Frost, Robert. "Mending Wall." The Poetry of Robert Frost: The Collected Poems, Complete and Unabridged. Holt, Henry & Company, Inc., 1955.
Godfrey-Smith, Peter. *Other Minds*. FSG, 2016.
Graziano, Michael. "How Phantom Limbs Explain Consciousness." *The Atlantic*. February 11, 2016.

Harvey, David. *The Condition of Postmodernity*. Blackwell, 1989.
Howe, Susan. *The Secret History of the Dividing Line*. Telephone Books, 1978.
Kim, Myung Mi. "[accumulation of land]." *Penury*. Omnidawn, 2009.
Kissinger, Henry. *World Order*. Penguin, 2014.
Latour, Bruno. *Down to Earth*. Polity Press, 2018.
Lorde, Audre. "After Images." The Collected Poems of Audre Lorde. W. W. Norton, 1997.
Luhmann, Niklas. *Ecological Communication*, trans. John Bednarz Jr. Polity Press, 1989.
Marsilius of Padua. *Defensor Pacis*, trans. Oliver J. Thatcher Internet Medieval Sourcebook, 1905.
Melville, Herman. "The Chapel." *Moby Dick*. Random House, 1981.
Moten, Fred and Harney, Stefano. *The Undercommons: Fugitive Planning & Black Study*. Autonomedia/Minor Compositions, 2013.
Si, Zhao. "Green Love." *Urban Poetry from China*, eds. Huang Fan, et al. Roof Books, 2021.
Von Uexküll, Jakob. *A Foray into the World of Animals and Humans*. University of Minnesota Press, 2010.
Winter, Jeffrey. *Oligarchy*. Cambridge University Press, 2011.
Wright, Richard. "Racial Bias in Negro Poetry." *Poetry*, March, 1944.

Syntax or How I Become What I Seem

Beltz, Adriene M. and Berenbaum, Sheri."How Early Hormones Shape Gender Development." *Curr Opin Behav Sci*. 2016 February; 7: 53–60. https://doi.org/10.1016/j.cobeha.2015.11.011.
Butler, Judith. *Gender Trouble*. Routledge, 1990.
Coolidge, Clark. *The Maintains*. This Press, 1974.
Darragh, Tina. "Just Lookin Just Lookin" *My Hands to Mutant Solidarities*. Dry Imager Press, 2020.
Darwin, Charles. *The Origin of Species*. Modern Library, 1998.
De Montaigne, Michel. "Of Experience." *Essays*. Penguin, 1993.
Donne, John. "The Canonization." www.poetryfoundation.org/poems/44097/the-canonization. 29 March 2020.
Fischer, Norman. "Emptiness And Compassion Go Hand In Hand." www.awakin.org/read/view.php?tid=2252. 18 May 2019.
Gilbert, Scott F. "A Symbiotic View of Life: We Have Never Been Individuals." *The Quarterly Review of Biology*, December 2012. https://doi.org/10.1086/668166. 24 July 2019.
Graziano, Michael A. Paraphrase from *Consciousness and the Social Brain*. Oxford University Press, 2013.
Iijima, Brenda. *Bionic Communality*. Roof, 2021.

Laitner, John., et al. "Incorporating Behavioural, Social, and Organizational Phenomena in the Assessment of Climate Change Mitigation Options." Jochem, et al. *Society, Behaviour, and Climate Change Mitigation*. Springer Netherlands, 2000.

McCaffery, Steve and bpNichol. *Rational Geomancy: The Kids of the Book-Machine. The Collected Research Reports of the Toronto Research Group 1973–1982*. Talon Books, 1992.

Nash, David George. *Topics in Walpiri Grammar*. www.ai.mit.edu/projects/dm/theses/nash80.pdf. 15 October 2021.

Pope, Alexander. "An Essay on Man". *The Poems of Alexander Pope*, ed. John Butt. Yale University Press, 1966. 507.

Rootsaction.org. 15 June 2018. act.rootsaction.org/o/6503/t/0/blastContent.jsp?email_blast_KEY=213802. 8 February 22. Also, Mohamad Bazzi. "The troubling overlap between Jared Kushner's business interests and US foreign policy." *The Guardian*. 8 July 2019.

Salah, Trish. *Lyric Sexology*. Roof, 2016.

Shantideva. www.tibethouse.jp/about/buddhism/text/pdfs/Bodhisattvas_way_English.pdf. 3 March 2020.

Thurston, Lisa M., et al. *Principles of Gender Specific Medicine*. Academic Press, 2017.

Torres, Edwin. *Quanundrum*. Roof, 2021.

Virgil. *The Aeneid*, trans. Robert Fagels. Penguin, 2006.

Whitman, Walt. "A Noiseless Patient Spider." *Leaves of Grass*. Library of America, 2011.

Social Groups Scale: Identities & Connections

Alexander, Will. *The Combustion Cycle*. Roof, 2021.

Baldwin, James. *Nobody Knows My Name*. Vintage Books, 1993.

Bloom, Peter. "Fight for your alienation: The Fantasy of Employability and the Ironic Struggle for Self-Exploitation." *Ephemera: Theory & Politics in Organizations*, 13(4) 2013.

Camus, Albert. *The Stranger*. Vintage Books, 1958.

Crazy Bull, Cheryl and Guillory, Justin. "Revolution in Higher Education." *Daedalus*. Vol. 147, No. 2, The MIT Press, Spring 2018.

Darwin, Charles. *The Descent of Man*. Penguin, 2004.

Detrow, Scott. "Democrats Blast Biden For Recalling 'Civil' Relationship With Segregationists." www.npr.org/2019/06/19/734103488/democrats-blast-biden-for-recalling-civil-relationship-with-segregationists. 13 September 2021.

Grosz, Elizabeth. *Becoming Undone: Darwinian Reflections on Life, Politics, and Art*. Duke University Press, 2011.

Hahn, Michael. *Proceedings of the National Academy of Sciences*. A study of 51 languages. www.pnas.org/content/117/5/2347. 1 April 2020.

Hofer, Matthew and Golston, Michael, eds. *The Language Letters*. New Mexico University Press, 2019.

Howe, Susan. *My Emily Dickinson*. New Directions, 2007.

Kropotkin, Peter. *Mutual Aid: A Factor of Evolution*. www.complementarycurrency.org/ccLibrary/Mutual_Aid-A_Factor_of_Evolution-Peter_Kropotkin.pdf

Li, Chen. "In the Corners of Our Lives." *Urban Poetry from China*, eds. Huang Fan and James Sherry. Roof Books, 2021.

Margulis, Lynn. *Journal of Theoretical Biology*. 14 (3): 255–274. https://doi.org/10.1016/0022-5193(67)90079-3. PMID 11541392, 8 February 2022.

Moretti, Franco. *Graphs Maps Trees*. Verso, 2007.

Morton, Timothy. "Spokesvultures for Ecological Awareness: an interview with Timothy Morton." https://galeapps.gale.com/apps/auth?userGroupName=&origURL=https%3A%2F%2Fgo.gale.com%2Fps%2Fi.do%3Fid%3DGALE%257CA496566676%26amp%3Bsid%3DgoogleScholar%26amp%3Bv%3D2.1%26amp%3Bit%3Dr%26amp%3Blinkaccess%3Dabs%26amp%3Bissn%3D03170802%26amp%3Bp%3DAONE%26amp%3Bsw%3Dw%26amp%3BuserGroupName%3Dnysl_oweb%26amp%3BisGeoAuthType%3Dtrue&prodId=GVRL. 14 October 2021.

Mosley, Walter. "Why I Quit the Writers' Room." *New York Times*. www.nytimes.com/2019/09/06/opinion/sunday/walter-mosley.html. 6 October 2019.

Nelson, Matthew J., et al. "Neurophysiological Dynamics of Phrase-Structure Building During Sentence Processing." www.pnas.org/content/pnas/114/18/E3669.full.pdf. 31 March 2020.

National Institutes of Health. NIH News Release. www.nih.gov/news-events/news-releases/nih-human-microbiome-project-defines-normal-bacterial-makeup-body. 13 June 2012.

Pasolini, Pier Paolo. *Saint Paul*. Verso, 2014.

Piketty, Thomas. *Capital in the 21st Century*. Belknap Press, 2017.

Rothenberg, David. *Survival of the Beautiful*, Bloomsbury Press, 2011.

Sahlins, Marshall and Service, Elman R., eds. *Evolution and Culture*. University of Michigan, 1988.

Schilpp, P. A., ed. "Evolutionary Epistemology." *The Philosophy of Karl R. Popper*. Open Court, 1974.

West, David. "How to Predict the Lifespan of a Company in One Simple Measure" www.itproportal.com/features/how-to-predict-the-lifespan-of-a-company-in-one-simple-measure/. 14 Februrary 2017.

Wilson, David Sloan. *This View of Life: Completing the Darwinian Revolution*. Pantheon, 2019.

Social Syntax

Aji, Hélène. "'Writing (as) (and) thinking': Charles Bernstein's Work 'in' Language." *Études anglaises* 2006/3 (Vol. 59).

Aristotle. *The Politics*, trans. Carnes Lord. University of Chicago Press, 2013.

Armantrout, Rae. "Why Don't More Women Do Language-Oriented Writing?" *L=A=N=G=U=A=G=E*, 1, eds. Bruce Andrews and Charles Bernstein. February, 1978.

Ballotopedia. "Wisconsin Collective Bargaining." ballotpedia.org/Wisconsin_collective_bargaining. 30 June 2021.

Boyd, Robert and Richerson, Peter. "Culture and the Evolution of Human Cooperation." The Royal Society Publishing, 2009. https://doi.org/10.1098/rstb.2009.0134.

Braudel, Ferdinand. *Civilization and Capitalism, 15th–18th Century.* Harper & Row, 1984. Vol 2.

Brockhurst, Michael A. "Sex, Death, and the Red Queen." *Science*, July 8, 2011.

Bulmer, Michael. "The Theory of Natural Selection of Alfred Russel Wallace FRS." The Royal Society Publishing, 2005. https://doi.org/10.1098/rsnr.2004.0081.

Burchman, Seymour. "How to Tie Executive Compensation to Sustainability." *Harvard Business Review*. https://hbr.org/2018/11/how-to-tie-executive-compensation-to-sustainability. 12 December 2018.

Coolidge, Clark. "Trilobite." *Space*. Harper and Row, 1970.

Carroll, Lewis. *Through the Looking Glass*. Createspace, 2017.

Croson, Rachel. "Theories of Commitment, Altruism and Reciprocity: Evidence from Linear Public Goods Games." *Economic Inquiry*, 45 (2), 199–216. 2007. https://doi.org/10.1111/j.1465-7295.2006.00006.x.

Darwin, Charles. *Descent of Man*. Global Grey eBooks, 2018. Based on Second edition 1874.

Environmental Defense Fund. www.edf.org/sites/default/files/Business-and-the-Fourth-Wave-of-Environmentalism_2019.pdf. 15 March 2020.

Folbre, Nancy. "Evolution and Intersectional Conflict." https://blogs.umass.edu/folbre/2021/01/18/evolution-and-intersectional-conflict/. 20 June 2021.

Goodwin, Brian. *How the Leopard Changed his Spots*. Scribner, 1994.

Hale, John. "Corporations Are Changing. Exhibit No. 1: Larry Fink's Letter." http://www.morningstar.com/articles/1075393/corporations-are-changing-exhibit-no-1-larry-finks-letter. 22 January 2022.

Hwang, Sung-Ha and Bowles, Samuel. *Journal of Economic Behavior & Organization*. Vol. 83, No. 3, August 2012. https://doi.org/10.1016/j.jebo.2012.06.001.

Inman, Peter. *Red Shift*. Roof, 1988.

Latour, Bruno. *Down to Earth: Politics in the New Climatic Regime.* Polity Press, 2013.

Lazer, Hank. *Opposing Poetries.* Northwestern University Press, 1996.

Melgard, Holly. *Fetal Position.* Roof Books, 2021.

Patterson, Hugh. "The Competitive Darwin." *Paleobiology*, 31(2), 2005. https://doi.org/10.1666/0094-8373(2005)031[0056:TCD]2.0.CO;2.

Perelman, Bob. "To My Mother." *The Future of Memory.* Roof, 1998.

Qianzi, Che. "Magnificent." *Urban Poetry from China.* Roof, 2021.

Silliman, Ron. "Realism: An Anthology of Language Writing." *Ironwood.* Ironwood Press, 1982.

Snyder, Gary. "Rip Rap." *Rip Rap and Cold Mountain Poems.* Shoemaker & Hoard, 2003.

The Drawdown Review. Project Drawdown, 2020. Drawdown.org. 20 February 2022.

Wilson, David Sloan. *Darwin's Cathedral.* University of Chicago Press, 2002.

How Can Culture Change Habitat?

Armantrout, Rae. "Bubble Wrap." *Money Shot.* Wesleyan University Press, 2011.

Armey Curve. thinkbynumbers.org/economics/gdp/ideal-level-of-government-spending/. 20 June 2015.

Bendell, Jem. "Deep Adaptation: A Map for Navigating Climate Tragedy" Revised 2nd Edition July 27, 2020. IFLAS Occasional Papers Volume 2. University of Cumbria, Ambleside, UK. insight.cumbria.ac.uk/id/eprint/4166/. 24 November 2021.

Berssenbrugge, Mei-mei. "Hello the Roses." *Hello the Roses.* New Directions, 2015.

Breuer, Pablo and Perlman, David. "Hacking Ten Million Useful Idiots: Online Propaganda as a Socio-Technical Security Project," www.youtube.com/watch?v=trxzyEfG8cY. Black Hat, 3 August 2019.

Hanisch, Carol. "The Personal Is Political." https://webhome.cs.uvic.ca/~mserra/AttachedFiles/PersonalPolitical.pdf. 13 February 2022.

Moore, Jason. *Capitalism and the Web of Life.* Verso, 2015.

Moten, Fred. *Black and Blur (consent not to be a single being).* Duke, 2017.

Reilly, Evelyn. "Song of." *Echolocation.* Roof, 2018.

Simon, Cecilia Capuzzi. *New York Times.* www.nytimes.com/2015/04/12/education/edlife/12edl-12mfa.html. 9 April 2015.

Stevens, Wallace. "Sunday Morning." *The Collected Poems.* Vintage, 2015.

Groups Build Social Syntax

Borges, Jorge Luis. "The Nothingness of Personality" (1922). *Selected Non-Fictions*. Viking, 1999.
Clark, Timothy. "Scale." *Telemorphosis: Theory in the Era of Climate Change*, vol. 1, ed. Tom Cohen. Open Humanities Press, 2012.
Corominas-Murtra, Bernat. "A Unified Approach to the Emergence of Complex Communication." Universitat Pompeu Fabra. digital.csic.es/bitstream/10261/46559/1/tbc.pdf. 13 February 2022.
Gorzelak, Monica A. et al. "Inter-Plant Communication Through Mycorrhizal Networks Mediates Complex Adaptive Behaviour in Plant Communities." *AoB Plants*. 2015; 7: plv050. www.ncbi.nlm.nih.gov/pmc/articles/PMC4497361/?msclkid=e2634700aba211ec9b64206090a9382a.
Jordan, June. "For the Sake of People's Poetry: Walt Whitman and the Rest of Us." *Passion: New Poems 1977–1980*. Beacon Press, 1980.
Keller, Lynn. "Twenty-First-Century Ecopoetry and the Scalar Challenges of the Anthropocene." *The News from Poems: Essays on the 21st-Century American Poetry of Engagement*, ed. Jeffrey Gray and Ann Keniston. University of Michigan Press, 2016.
Kim, Hyunju et al. "Universal scaling across biochemical networks on Earth." *Science Advances*. 16 January 2019: Vol. 5, No. 1, eaau0149. https://doi.org/10.1126/sciadv.aau0149.
Moore, Jason. *Capitalism and the Web of Life*. Verso, 2015.
Smil, Vaclav. *The Earth's Biosphere*. MIT Press, 2003.
Steels, Luc. "Modeling the cultural evolution of language" *Physics of Life Reviews* 8. 2011.
Vernadsky, Vladimir. *The Biosphere*. Copernicus, 1998.
Von Humboldt, Alexander. *Cosmos*. Sagwan Press, 2018.
Whitman, Walt. "Preface." *Leaves of Grass*. American Renaissance, 2009.
Wilson, David Sloan. *Darwin's Cathedral*. University of Chicago Press, 2002.
Woods, Derek. "Scale Critique for the Anthropocene." *Minnesota Review*, Issue 83, 2014. www.academia.edu/6727270/Scale_Critique_for_the_Anthropocene?msclkid=435e183aaba311ec8e15f563bc546869.
Woods, Derek. "Terminal Poetics of the Anthropocene." *The Cambridge Companion to 21st Century American Poetry*, ed. Timothy Yu. Cambridge University Press, 2021.

Some Structures of Syntax—Parataxis & Hypotaxis

Ashbery, John. *Girls on the Run*. FSG, 1999.
Che, Qianzi. "A Song." tr. Jeff Twitchell. http://jacketmagazine.com/20/pt-chinese.html. 15 November 2019.

Hejinian, Lyn. "Variations: A Return of Words." *A Thought is the Bride of What Thinking*. Tuumba, 1976.
Moten, Fred and Harney, Stefano. *The Undercommons: Fugitive Planning & Black Study*. Autonomedia/Minor Compositions, 2013.
Perelman, Bob. *The Marginalization of Poetry: Language Writing and Literary History*. Princeton University Press, 1996.
Rankine, Claudia. *Citizen*. Greywolf Press, 2014.
Salah, Trish. *Lyric Sexology Vol. 1*. Roof Books, 2014.
Sherry, James. "Snowball Earth Creates Innovative Poetry." *Oops!: Environmental Poetics*. BlazeVox, 2013.
Silliman, Ron. *The Age of Huts*. University of California Press, 2007.
Sun, Dong and Sherry, James. *The Reciprocal Translation Project*. Roof Books, 2018.

Scalable Syntax: Poetry Model of the Biosphere

Carr, E. Summerson and Lempert, Michael. *Scale: Discourse and Dimensions of Social Life*. University of California Press, 2016.
Foucault, Michel. *Politics, Philosophy, Culture*. Routledge, 1990.
Gampel, Yolanda. "The Interminable Uncanniness." *Psychoanalysis at the political border*, eds. Leo Rangell and Rena Moses-Hrushovski. International Universities Press, 1996.
Hannan, Michael T. and Freedman, John. "The Population Ecology of Organizations." *The American Journal of Sociology*, vol. 82, no. 5. March 1977. https://environment-ecology.com/history-of-ecology/132-history-of-ecology.html. 13 February 2022.
Kim, Hyunju, et al. "Universal scaling across biochemical networks on Earth." *Science Advances*, Vol. 5, No. 1. 16 January 2019. https://doi.org/10.1126/sciadv.aau0149
Merchant, Carolyn. *The Death of Nature: Women, Ecology, and the Scientific Revolution*. HarperOne, 1990.
Rowlands, Mark. "Environmental Epistemology." *Ethics & the Environment*, 10 (2), 2005.
Shelley, Percy Bysshe. "A Defense of Poetry." *A Defense of Poetry and Other Essays*. Kindle Edition, 2004.
Sherry, James. "Snowball Earth Creates Innovative Poetry." *Oops! Environmental Poetics*. BlazeVox, 2013.
Weinberger, Elliot and Paz, Octavio. *19 Ways of Looking at Wang-Wei*. New Directions, 2016.

The Anthropocene: Ecosystems & Time Frames

De Castro, Eduardo Viveiros and Danowski, Deborah. "Humans and Terrans in the Gaia War." *A World of Many Worlds*. Duke, 2018. https://doi.org/10.1215/9781478004318-007.

Deleuze, Giles & Guattari, Felix. *What Is Philosophy?* Verso, 2003.

Lewis, Simon L. and Maslin, Mark A. "Defining the Anthropocene." *Nature* 519. March 12, 2015.

Luhmann, Nicklas. *Ecological Communication*. Tr. John Bednarz. University of Chicago, 1989.

Moore, Jason. *Capitalism and the Web of Life*. Verso, 2015.

Morgan, Roy. www.roymorgan.com/findings/8044-most-important-problems-facing-australia-world-july-2019-201907020950. 11 August 2019.

Ruddiman, William. *Plows, Plagues, and Petroleum: How Humans Took Control of Climate*. Princeton, 2016.

Skinner, Jonathan. "Call the Pulsing Home." *Ecocriticism: Environments in Anglophone Literatures*, ed. Sonja Frenzel and Birgit Neumann. Anglistik and englischunterricht, 2017.

Von Uexküll, Jakob. *A Foray into the World of Animals and Humans*. University of Minnesota Press, 2010.

Vernadsky, Vladimir. *The Biosphere*. Copernicus, 1998.

Woods, Derek. "Scale Critique for the Anthropocene." *Minnesota Review*, Issue 83. Duke University Press, 2014.

Networks of Metaphor

Ahrens, Kathleen, et al. "Functional MRI of Conventional and Anomalous Metaphors in Mandarin Chinese." https://doi.org/10.1016/j.bandl.2005.10.004.

Armantrout, Rae. "Close." *Entanglements*. Wesleyan, 2017.

Arzouan, Yossi, et al. "Dynamics of Hemispheric Activity During Metaphor Comprehension: Electrophysiological Measures." https://doi.org/10.1016/j.neuroimage.2007.02.015.

Baldamus, W. "Networks." *The Sociology of Wilhelm Baldamus: Paradox and Inference* ed. M. Erickson, M. and C. Turner, C. Ashgate Publishing, 2010 (1982).

Bartlett, Frederick C. "Imagery in Thought." *Galileo's Commandment*, ed. Edmund Blair Bolles. W.H. Freeman, 1997.

Erickson, Mark. "Network as Metaphor." *International Journal of Criminology and Sociological Theory*. Vol. 5, No. 2 (2012). 912. https://ijcst.journals.yorku.ca/index.php/ijcst/article/view/35701. Accessed 12 June 2022.

Goatly, Andrew. *The Language of Metaphors*. Routledge, 2011.

Goffman, Erving. *Frame Analysis: An Essay on the Organization of Experience.* Harper and Row, 1974.
Graziano, Michael A. *Consciousness and the Social Brain.* Oxford University Press, 2015.
Harvey, David. "The Nature of Environment: The Dialectics of Social and Environmental Changes." *The Socialist Register*. https://cialistregister.com/index.php/srv/issue/view/427.
Hejinian, Lyn. *The Unfollowing 77.* Omnidawn, 2016.
kaufman, erica. *Post Classic.* Roof, 2019.
Lacan, Jacques. *Ecrits*. W. W. Norton, 2002.
Lakoff, George & Johnson, Mark. *Metaphors We Live By*. University of Chicago Press, 1980.
Ranasinghe, Champika Epa & Kray, Christian. www.researchgate.net/figure/Signal-propagation-reflection-refraction-and-diffraction_fig3_329009261. 13 February 2022.
Reddy, Michael J. "The Conduit Metaphor: A Case of Frame Conflict in Our Language about Language." *Metaphor and Thought*, ed. Andrew Ortony. Cambridge University Press, 1993.
Schön, Donald. "Generative Metaphor: A Perspective on Problem-Setting in Social Policy." *Metaphor and Thought*, ed. Andrew Ortony. Cambridge University Press, 1993.
Sherry, James *Oops! Environmental Poetics*. BlazeVOX, 2013.
Tenenbaum et al., "How to Grow a Mind: Statistics, Structure, and Abstraction." *Science*, 11 March 2011: Vol. 331, Issue 6022.
The Five Graces Group. "Language is a Complex Adaptive System." Language Research Learning Club, University of Michigan, 2009. https://doi.org/10.1111/j.1467-9922.2009.00533.x
Von Uexküll, Jakob. *A Foray into the World of Animals and Humans*. University of Minnesota Press, 2010.

Environmental Identities

Barad, Karen. *Meeting the Universe Halfway*. Duke University Press, 2007.
Bendell, Jem. "Deep Adaptation: A Map for Navigating Climate Tragedy Revised." https://jembendell.com/2019/05/15/deep-adaptation-versions/. 13 February 2022.
Coolidge, Clark. *The Book of During*. The Figures, 1991.
Deleuze, Giles. *Difference and Repetition*, trans. Paul Patton. Columbia, 1994.
Höfert, Almut. *Describing the Enemy*. Campus Historiche Studien, 2003.
Hofer, Matthew and Golston, Michael, eds. *The Language Letters: Selected 1970s Correspondence of Bruce Andrews, Charles Bernstein, and Ron Silliman*. University of New Mexico Press, 2019.

Parsons, Talcott. "Social Structure and the Development of Personality." *Psychiatry*, 21:4, 1958. https://doi.org/10.1080/00332747.1958.11023143.
Stein, Gertrude. *How to Write*. Dover, 1975.
Thompson, Michael. "Man and Nature as a Single but Complex System." *Encyclopedia of Global Environmental Change. Vol. 5: Social and Economic Dimensions of Global Environmental Change*, ed. Peter Timmerman. John Wiley & Sons Inc., 2002.
Todes, Daniel P. *Darwin Without Malthus: The Struggle for Existence in Russian Evolutionary Thought*. Oxford, 1989.

Environmental Autobiography

Armantrout, Rae. "Scripture." *Just Saying*. Wesleyan, 2013.
Armantrout, Rae. "Xenophobia." *Extremities*. The Figures, 1978.
Fink, Thomas. "Interview with James Sherry." https://dichtungyammer.wordpress.com/. 12 December 2021.
Najmabadi, Afsaneh. *Professing Selves*. Duke University Press, 2014.
Patokorpi, Erkki. "Information Pluralism and Some Informative Modes of Ignorance." *Information*. 2011, 2. https://doi.org/10.3390/info2010041.

Social Syntax II: Linkages & Connectors

Armantrout, Rae, et al. *The Grand Piano. Mode A*, ed. Barrett Watten. This Press, 2007.
Chetty, Raj, et al., "Where Is the Land of Opportunity? The Geography of Intergenerational Mobility in the United States." *The Quarterly Journal of Economics.* June, 2014. https://academic.oup.com/qje/article/129/4/1553/1853754. 20 November 2021.
Gillis, Justin. "Flooding of Coast, Caused by Global Warming, Has Already Begun." *New York Times*, 4 September 2016.
Golston, Michael and Hofer, Matthew, eds. *The Language Letters: Selected 1970s Correspondence of Bruce Andrews, Charles Bernstein, and Ron Silliman.* University of New Mexico Press, 2019.
Grenier, Bob. "On Speech." *This* 1. 1971.
Piketty, Thomas. *Capital in the 21st Century*. Belknap Press, 2017.

Identity's Constructive & Connective Ecosystem

Bloch, Maurice. *Going In and Out of Each Other's Bodies*. Routledge, 2012.
Bloch, Maurice. "The Resurrection of the House Amongst the Zafimaniry of Madagascar." *About the House: Lévi-Strauss and Beyond*, ed. J. Carsten and S. Hugh-Jones. Cambridge University Press, 1995.

Cohen, Cathy and Jackson, Sarah. "Ask a Feminist: A Conversation with Cathy Cohen on Black Lives Matter, Feminism, and Contemporary Activism." https://signsjournal.org/ask-a-feminist-cohen-jackson/. 17 March 2020.

Crenshaw, Kimberlé. *On Intersectionality: Essential Writings*. The New Press, 2022.

El Gamal, Rania and Carvalho, Stanley. "Saudi Arabia Sees Domestic Energy Use Falling, Plans Renewables Push." Reuters. www.reuters.com/article/us-saudi-energy-reforms/saudi-arabia-sees-domestic-energy-use-falling-plans-renewables-push-idUSKCN1P918N. 15 January 2019.

Feyerabend, Paul. *Against Method*. Verso, 2010.

Graziano, Michael A. *Consciousness and the Social Brain*. Oxford University Press, 2015.

Grenier, Bob. "Northern New Hampshire." *This Magazine*. No. 7, ed. Barrett Watten. Spring 1975.

Haraway, Donna. *When Species Meet*. University of Minnesota Press, 2008.

Hartley, George. *Textual Politics and the Language Poets*. Indiana University Press, 1989.

hooks, bell. *Feminist Theory: From Margin to Center*. Routledge, 1984.

kaufman, erica. *Instant Classic*. Roof, 2014.

Ma, Jun. https://time.com/5669061/china-climate-change/. 7 July 2020.

Marcos, Subcommandante. *Our Word Is Our Weapon*. Seven Stories Press, 2004.

McLuhan, Marshall. *Understanding Media*. MIT Press, 1994.

Mead, George Herbert. *The Philosophy of Act*. University of Chicago Press, 1938.

Olson, Charles. *Maximus*, III. University of California Press, 1985.

Pound, Scott. "The Chinese Notebook and the Materialities of Communication." http://vectors.usc.edu/thoughtmesh/publish/191.php#. 7 July 2018.

Premack, David and Woodruff, Guy. "Does the Chimpanzee Have a Theory of Mind?" *Behavior and Brain Sciences*. Cambridge University Press, 1978.

Reddy, Michael J. "The Conduit Metaphor: The Case of Frame Conflict in our Language about Language." *Metaphor and Thought*. Cambridge University Press, 1993.

Salwiczek, Lucie, et.al. "What Do Animals Remember about Their Past?" *Learning and Memory: A Comprehensive Reference*. 2008. https://doi.org/10.1016/B978-012370509-9.00060-7.

Sherry, James and Gottlieb, Michael, eds. *Roof Magazine*. #4: Washington, DC. #5: San Francisco. eclipsearchive.org/projects/ROOF/roof.html. 20 October 2021.

Skinner, Jonathan. "Conceptualizing the Field." https://jacket2.org/commentary/jonathan-skinner. 8 July 2020.

Stanford Encyclopedia of Philosophy. Common Theories of the Organization of Organisms Are Summarized Here. https://plato.stanford.edu/entries/levels-org-biology/. 2 March 2022.

Tenenbaum, Joshua. et al. "How to Grow a Mind: Statistics, Structure, and Abstraction." *Science*, 11 March 2011: Vol. 331, Issue 6022.
UNESCO. https://ich.unesco.org/en/RL/woodcrafting-knowledge-of-the-zafimaniry-00080. 13 July 2019.

The Condition & Hierarchy of Identification

Atkin, Ron. *Multidimensional Man*. Penguin, 1982.
Dworkin, Craig. *Motes*. Roof, 2011.
Gould, Stephen Jay. *The Structure of Evolutionary Theory*. Belnap, 2002.
Grant, Peter and Rosemary. *40 Years of Evolution: Darwin's Finches on Daphne Major Island* Princeton University Press, 2014.
Shelley, Percy Bysshe. "Mutability." *John Keats and Percy Bysshe Shelley Complete Poetical Works*. The Modern Library, 1940.
Tardos, Anne. *Nine*. BlazeVox, 2015.
Templeton, Fiona. *Cells of Release*. Roof, 1997.
Toscano, Rodrigo. *Explosion Rocks Springfield*. Fence, 2016.

Impact of Text & Environment on Another Self

Boughn, Michael and Johnson, Kent, eds. "Poetry Wars, 1984—Documents from *Poetry Flash*, including 'Of Duncan and Zukofsky on Film: Traces Now and Then' by David Levi Strauss." *Dispatches from the Poetry Wars*. https://wayback.archive-it.org/www.dispatchespoetrywars.com/documents/poetry-wars-1978-documents-from-the-poetry-flash-including-of-dunncan-and-zukofsky-on-film-traces-now-and-then-by-david-levi-strauss/. 12 February 2022.
Caso, Juan Manuel Rodríguez. "Darwin and Wallace on Morals and Ethics: Two Different Views from Natural Selection." Universidad Autonoma de Mexico. www.academia.edu/1675526/Darwin_and_Wallace_on_Morals_and_Ethics_Two_Different_Views_from_Natural_Selection. 13 February 2022.
Eldakar, Omar Tonsi and Wilson, David Sloan. "Selfishness as Second-order Altruism." Proc Natl Acad Sci U S A. May 13, 2008. https://doi.org/10.1073/pnas.0712173105. 12 February 2022.
Kramer, Jos and Meunier, Joël. "Kin and Multilevel Selection in Social Evolution: A Never-Ending Controversy?." https://doi.org/10.12688/f1000research.8018.1. 28 September 2019.
Levi-Strauss, Claude. "Translators Preface." *Structural Anthropology*. Basic Books, 2008.
Maxwell, James Clerk. "Molecular Evolution." www.poetryfoundation.org/poems/45777/molecular-evolution. 12 February 2022.
Sherman, Alexie. "Evolution." www.poemhunter.com/poem/evolution-4/. 12 February 2022.

Index

J. Sherry, *Selfie*, https://doi.org/10.1007/978-981-19-4870-1

CPSIA information can be obtained
at www.ICGtesting.com
Printed in the USA
LVHW020902140423
744372LV00003B/326

9 789811 948695